Advanced Photovoltaic System Design

John Balfour, MEP, PhD,
LEED Accredited Professional

Michael Shaw, MBA, PhD

Nicole Bremer Nash, BA

World Headquarters
Jones & Bartlett Learning
5 Wall Street
Burlington, MA 01803
978-443-5000
info@jblearning.com
www.jblearning.com

Jones & Bartlett Learning books and products are available through most bookstores and online booksellers. To contact Jones & Bartlett Learning directly, call 800-832-0034, fax 978-443-8000, or visit our website, www.jblearning.com.

Substantial discounts on bulk quantities of Jones & Bartlett Learning publications are available to corporations, professional associations, and other qualified organizations. For details and specific discount information, contact the special sales department at Jones & Bartlett Learning via the above contact information or send an email to specialsales@jblearning.com.

This book contains material reprinted with permission from NFPA 70®, *National Electrical Code®*, Copyright © 2010, National Fire Protection Association, Quincy, MA. This reprinted material is not the complete and official position of the NFPA on the referenced subject, which is represented only by the standard in its entirety.

Production Credits

Chief Executive Officer: Ty Field
President: James Homer
SVP, Chief Technology Officer: Dean Fossella
SVP, Chief Marketing Officer: Alison M. Pendergast
SVP, Curriculum Solutions: Christopher Will
VP, Design and Production: Anne Spencer
VP, Manufacturing and Inventory Control: Therese Connell
Editorial Management: High Stakes Writing, LLC, Editor
  and Publisher: Lawrence J. Goodrich
Managing Editor, HSW: Ruth Walker
Reprints and Special Projects Manager: Susan Schultz
Senior Marketing Manager: Andrea DeFronzo

Marketing Manager: Lindsay White
Manufacturing and Inventory Control Supervisor: Amy Bacus
Composition: Abella Publishing Services
Cover Design: Kristin E. Parker
Rights and Permissions Manager: Katherine Crighton
Photo Research Supervisor: Anna Genoese
Cover Image: © DSBfoto/ShutterStock, Inc.
Chapter Opener and Notes Image: © Visional/Fotolia.com
Tech Tip Image: Courtesy of NASA
Printing and Binding: Malloy, Inc.
Cover Printing: Malloy, Inc.

ISBN: 978-1-4496-2469-9

Library of Congress Cataloging-in-Publication Data
Unavailable at time of publication.

6048
Printed in the United States of America
15 14 13 12 11   10 9 8 7 6 5 4 3 2 1

Brief Contents

Contents

Chapter 2 Evaluation and Design Criteria Part II 23

Preface

THE PHOTOVOLTAICS (PV) INDUSTRY stands on the brink of a revolution. The appeal of a new and growing industry has brought an influx of new PV professionals to the market, but the availability of educational resources has not kept pace with market demands. This gap has led to serious quality and performance issues that the industry will need to face in the decades ahead.

The *Art and Science of Photovoltaics* series was developed to fill this education gap. Each book in the series goes beyond simple systematic processes by tackling performance challenges using a systems perspective. Readers do not learn PV design and installation steps in a vacuum; instead they gain the knowledge and expertise to understand interrelationships and discover new ways to improve their own systems and positively contribute to the industry.

This series is aimed at both the novice and expert. The texts take the learner from an overview of PV, through simple design and installation steps and considerations, to the design and installation of high-performance systems. The series also prepares readers for the NABCEP PV Entry Level Exam.

Readers will come to understand PV using a system perspective. This includes how the selection and installation of each component in the system influences other components. Together, each design and installation consideration has a considerable impact on the ability of the PV system to generate power efficiently over its lifetime. The series also focuses on high-performance PV—what it is, how it can be designed and installed, and how performance improves the client-customer relationship and the industry as a whole.

Advanced Photovoltaic System Design provides readers with the ability to understand, design, and recognize high-performance PV systems. The single most important concept that readers will gain is that small steps and decisions add up incrementally to sizeable energy production increases, longer system and component lives, and less maintenance costs. More than anything else, performance PV is about producing more electricity for less cost. Readers will also learn the difference between designing for code and designing for performance. Readers will complete the text with the ability to create basic three-line PV diagrams.

The writers and editors of the *Art and Science of Photovoltaics* series sincerely hope it plays a role, not only in the reader's success in a PV-related career, but in raising the standards and professionalism of the industry as a whole. This, it is hoped, will contribute to the nation's energy security and to a cleaner and less carbon-dependent environment.

Overview of Advanced Photovoltaic (PV) System Design and Design Criteria

TO UNDERSTAND ADVANCED PHOTOVOLTAIC (PV) system design, you must look at PV not just as a system, but as a system of systems. Each PV system design is unique. Components make up each system. In all, designing PV systems involves 36 criteria, categorized into six groups. Each group relates to the others like cogs in a wheel. As a whole, they form a system. This system revolves around the goals of the client.

Topics & Concepts

This chapter covers the following topics and concepts:

- Grades of PV systems
- Understanding positions in the PV field
- Sales, client requirements, goals, and objectives
- Usage and usage profile
- Budget (system size and economic viability analysis)
- Training/education/staging/ delivery
- Solar resource
- Location
- Climate
- Altitude
- Shading
- Airflow
- Wind
- Space available
- Structural challenges
- Orientation
- Mounting
- Maintenance of the structure on which the system is mounted
- Aesthetics
- System life/system service life
- Warranty
- Utility and energy production rate structure

Goals

Upon completion of this chapter, the student will be able to:

- Understand the relationships of the members of the PV system team and how they relate to the sales process
- Understand the relationship between the earth and sun
- Have a basic understanding of 20 of the 36 criteria that make up a PV system

The six groups of criteria are as follows:

- **Client**—The client is the most important part of a system design. The client dictates the goals and objectives used to drive the entire design process. The designer must compare these goals with what is possible and manage client expectations carefully. The criteria in the client group include the following:
    - Sales, client requirements, goals, and objectives
    - Usage and usage profile
    - Budget (system size and economic viability analysis)
    - Monitoring
    - Operations and system maintenance
    - Cleaning (panels, filters, and equipment cavities)

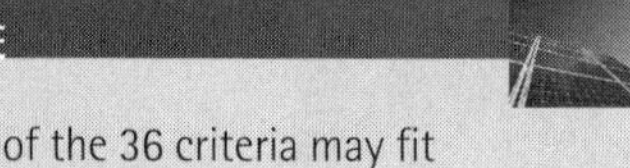

Many of the 36 criteria may fit into more than one group.

- **Designer and installer**—The designer and installer represent the hub of the wheel. They must balance the interests of the other components in the system while meeting the goals of the client. Criteria in this group include the following:
    - Training/education/staging/delivery
    - System performance modeling
    - Quality of installation
    - Testing schedule and procedures
    - Quality assurance (continual reevaluation and review)
    - Equipment burn-in response

- **Environment**—A system has little value if does not have access to adequate solar resources. Climate affects how you build the system and the longevity of the system. Criteria in this group include the following:
    - Solar resource
    - Location
    - Climate
    - Altitude

- Shading
- Airflow
- Wind

- **Structure**—The structure plays a key role in what you can build and how you will build it. The absence of a suitable mounting structure means that the PV designer must create and locate one. This group includes the following criteria:
  - Space available
  - Structural challenges
  - Orientation
  - Mounting
  - Maintenance of the structure on which the system is mounted
  - Aesthetics
  - Service entrance section (SES)

- **Government**—This includes any organization or body that may dictate how, when, and where you can build a PV system. It also includes regulations on selling power back to a utility. Criteria in this group include the following:
  - National electrical codes
  - Local codes

- **Manufacturers**—Manufacturers create the components, design parameters, and warranties. Their ability to design and test components directly influences the effectiveness of the system. Criteria listed in this group include the following:
  - System life/system service life
  - Warranty
  - Panels
  - Panel certification by Underwriters Laboratories (UL), the German organization known as TÜV, and other testing bodies
  - Strings, string sizing, and string placement
  - Inverter(s)
  - Balance of system
  - Wire run and sizing (impacts conduit), wire run locations, protection

> **NOTE**
>
> Currently, the PV industry maintains no body of standards for designers and installers to follow. Installers do, however, need to obtain a copy of local building codes and permits before installation begins. In addition, the PV designer, in the United States, must follow the National Electrical Code (NEC) and a local safety code.

Experts have developed these criteria during the course of decades of experience in the PV industry. Chapters 1 and 2 review the 36 criteria as they occur in actual design scenarios. These chapters include details of what works—and how to avoid what does not. This chapter explains the first 20 of the 36 criteria. It begins with information about selling a PV system. It ends with the operations and maintenance of a system.

Grades of PV Systems

There are three grades of PV systems:

- Subprime
- Moderate
- High performance

Within a few years, subprime PV companies and systems will disappear. New industry standards and regulations will force out their low-quality work and products.

High-performance systems provide more kilowatt-hours per dollar spent. They have lower lifetime operation costs, and have longer functional lives. High-performance systems are a much better value for the client. You can use the 36 criteria as a road map to designing and installing high-performance systems.

Understanding Positions in the PV Field

It is a team effort to successfully sell, design, install, and service a PV system. The following definitions should help you to understand the different roles in a PV team:

- **Sales and marketing**—The PV sales and marketing staff is composed of the individual or team that initiates and completes an agreement (contract) with the client to deliver a properly designed and working PV system. The sales and marketing team typically works for the PV integrator (see below).
- **PV integrator**—The PV integrator is usually a single company. The integrator determines the level of work and quality of the system. It makes the sale, then designs and installs the system. The integrator provides all support services during the life of the PV system.
- **PV designer**—The PV designer provides design services. The designer can be an individual, organization, or team. Designers research the site and identify equipment needed. They also perform calculations, and evaluate and analyze local environmental conditions. Services include providing design drawings. A good PV designer designs moderate- or high-performance PV systems.
- **PV installer**—PV installers are professionally trained. They assemble, test, and support the PV system.

The integrator and client must establish, develop, and agree upon clear and concise system-design goals and objectives. This is important to the PV system's delivery and performance. Regardless of how the PV team is organized, it must address the 36 criteria.

Sales, Client Requirements, Goals, and Objectives

Selling and designing a PV system for a client is a team effort. Every person who encounters the client is responsible for selling the PV system. That includes office personnel. The salesperson, however, usually has the first contact with the client.

Following are selling tips to help PV professionals make the sale:

- Establish trust with the potential client. Make the client feel comfortable, and feel that the company has the client's best interest in mind.
- Be sincere with each client. Treat clients with respect.
- Believe in your product. If you believe, the client will believe.
- Communicate with the client. Understand why the client wants a PV system. Help fulfill those desires.
- Give the client tips on how he or she can save energy and money.
- Build value into the system before giving prices. This will help clients understand the value they will receive for their money.
- Not everyone will purchase a system. If you make a great impression, however, people will refer your company to friends and family looking for a PV system. And of course, some clients who do not purchase immediately may purchase in the future.
- A satisfied client is the best advertisement a company can have.

Remember, client satisfaction is the No. 1 priority.

Usage and Usage Profile

No two properties use energy in exactly the same way. Creating a usage profile, also called a load profile, will help the PV designer understand how the client uses energy.

To create a usage profile, it is important to do the following:

- List, evaluate, and understand the client's current and future energy usage patterns.
- Educate the client on how to modify energy usage to get the greatest economic gain from the PV system.
- Coach the client on how to reduce electrical energy usage. Explain how electrical rates versus cost of the PV system will affect the client over the life of the PV system.
- Understand and be able to demonstrate ways the client can reduce or eliminate energy demands, thus reducing the size and cost of the PV system.

- Pay particular attention to changes in seasonal patterns as compared to energy-production patterns. Understand them and use them to the client's benefit.

Budget (System Size and Economic Viability Analysis)

Costs and budget constraints drive most PV systems. Indeed, budget will drive the PV system strategy, size, phasing, and overall outcome. It is the responsibility of the PV system sales team and designer to understand and explain to the client how costs, incentives, and financing will affect the final PV system.

If the integrator opts for high-performance system design and delivery, it is critical that the integrator clearly and accurately outline for the client the impact of that decision on budget.

To understand how the client's budget will affect the size of the PV system, do the following:

- Collect data on and understand how budget conditions will affect the final PV system. These budget conditions include cash, the client's ability to finance, incentives, lease options, phasing options, and changes in usage patterns over time.
- Compare real local system-monitoring data and determine real output of the final working PV system.
- Provide a clear analysis that demonstrates the superiority of the number of kilowatt-hours per dollar over the cost of DC watts as a metric of cost-effectiveness. The better value is in kilowatt-hours. The difference in output between a high-performance PV system and one that performs only moderately well can be 10 to 25 percent.
- Explain these points to the client so the client can make informed decisions during the PV system design and installation phases.

Training/Education/Staging/Delivery

There are three steps in mastering photovoltaic technology and becoming a seasoned solar professional:

- Become an apprentice.
- Become a journeyman.
- Become a master-level professional.

A solar-integration company that plans to be in business for a long time will have to invest in staff. This includes training and cross-training staff to work toward continual improvement. Placing poorly trained, underskilled staff on a job risks the safety and health of both the workers and the client. It also results in low-quality systems. Installing subprime systems undermines the integrator and the industry. It's not ethical to send out employees who do not have the ability to deliver a good product. What's more, doing so may put the integrator out of business.

No matter what level the PV system professional is, he or she always needs to be aware of the importance of the following:

- Training is a never-ending process of learning how to maximize professional office and job-site skills. It is a function usually driven by the company and its management. Training often occurs on the job, under the supervision of a more skilled and knowledgeable individual or team.
- The solar professional drives the education process, not the company. Education requires personal time for reading, attending professional classes, networking, and the continual seeking of professional self-improvement. An employee who is willing to invest in his or her education is a solar professional and will be an inspiration to the team.
- Properly staging a project requires a complete understanding and comprehension of systems, design, installation, and the order in which tasks should be completed, as well as the provision of material, labor, and all resources to properly complete a job. It is the flow of all component parts from the first client meeting though warranty.
- Delivery is the process of providing a client with a good-quality working PV system, installed with quality workmanship that meets or exceeds the client's expectations.

Solar Resource

The sun continually provides the earth with energy in the form of solar radiation. That energy gives heat and light to plants and animals. If just 0.01 percent of the sun's energy were collected every day, it would replace all the energy produced using uranium, gas, coal, and oil combined.

To help reduce dependence on those forms of energy, PV systems collect solar energy in a device called a **panel** or **module**. The solar panel captures solar radiation as direct current (DC), and an **inverter** converts it to alternating current (AC). Panels assembled into **arrays** collect enough energy to power appliances, tools, illuminated road signs, factories, and many other things. New PV applications are constantly being created.

Concepts the PV professional should understand about Earth's relationship to the sun are as follows:

- All local environmental conditions affect the local solar resource and drive energy production. Environmental conditions include weather and temperature, as well as buildings, mountains, and wind.

- The PV system's performance output depends on the level and quality of detail put into the system's design and installation. The more specific the detail put into the system by the PV designer and carried out by the installation team, the better the PV system will perform.
- The integrator must understand all phases of the sun's physical relationship to the PV panel surface.
- Using a computer model for the PV system will help the client understand the necessity of each component of the PV system. This includes images of systems, potential layouts for equipment, energy and financial analyses, and operations and maintenance details and costs.

Location

Location affects design and performance of the PV system. Observe, evaluate, and consider all local factors as they exist when designing the PV system. Add considerations for the client's future energy needs. Many integrators and designers reduce their systems to one or two simple variations or options. This strategy may save them money in the short term, but the costs to the consumer are high. Designers must take a comprehensive look at all the location factors.

Location factors that you should consider when designing a PV system include the following:

- Understand where equipment will be located and possible obstructions that may interfere with the PV system's performance.
- Understand the impact and relationship of the system's location with the complete local environment.
- Evaluate and analyze all location factors such as shading from structures, antennas, trees, mountains, and future shading issues.
- Location influences all design factors. For instance, the system may need extra support due to high winds. Ice and snow may be a problem, or the growth of nearby trees may increase shading in the future.

Climate

Climate affects PV system output. You must collect data on local temperature, rain, clouds, freezing, and other microclimate issues from a reputable source. You must then understand the ramifications of this data on the PV system. Local data is essential to good PV design and system performance. Be aware that string sizing and system modeling must be carefully adjusted for real climate conditions.

Here are points to keep in mind about how climate affects the PV system:

- You must have a complete understanding of how panels and inverters work at a variety of temperatures and climatic conditions.

- You must understand the impact of climate extremes on performance and equipment longevity.
- You must understand that standard test conditions (STC) and **performance test conditions (PTC)** do not represent actual climate conditions. The PV designer must design for the conditions in which the system will operate. Designers should obtain the skill to reflect real conditions in their designs and analyses.
- Study the climatic data and, when possible, develop corroborating information from local professionals as to what the irregularities are.
- Measure summer and winter panel and cell temperatures. Most design tools use ambient temperatures. These do not reflect real temperatures as measured on the cell. Cell temperature drives the system performance.

Altitude

Consider altitude when gauging real operational conditions. Altitude can significantly affect the efficiency of the PV system. Points to consider include the following:

- Altitude has an impact on all local environmental factors.
- Altitude will have an impact on cooling and heating for both panels and inverters.
- Differences in altitude result in different spectral values for systems.
- Higher-altitude conditions can enhance performance.
- Higher-altitude conditions can result in the system's ability to collect additional energy due to lower atmospheric conditions.
- Higher-altitude conditions can result in quick cooling at night and faster heating of module surfaces in the day.

Shading

You must understand the consequences of shading and be able to communicate them to the client. There are many tools on the market to assist the PV professional in determining shading issues. One such tool is the SunEye.

The PV integrator must understand how shade-determining tools function. These tools are only as accurate as the data put into the program. You must pay attention to the whole site. You must also anticipate future changes. For example, during the design process, you must consider and discuss new trees, buildings, mountains, and other objects that can cast shadows from hundreds of feet away. Information gathered during this part of the process will save the PV designer and the installer time in the long run.

NREL engineer James Salasovich checks his SunEye, a device that analyzes shade to help determine a site's solar-energy potential. The SunEye can distinguish among trees, clouds, and shaded contours.

Courtesy of US Department of Energy/National Renewable Energy Laboratory (DOE/NREL): Credit: Patrick Corkery

Following are several points about shading that the integrator should understand:

- The design and installation must consider in detail the impact of self-shading—that is, shading of modules by other modules or components.
- Many designers fail to understand micro inverter values in dealing with shading issues. This has a negative impact on the overall system performance. Many times, using micro inverters lowers the system performance.
- Seasonal shading is caused by obstructions that are north, east, and west of the panel, from equinox to equinox in the spring and summer.
- Orientation shifts from south to north when you work in the Southern Hemisphere.

Shading of cells and panels has a serious impact on the system. It is important to gain the knowledge and skill necessary to recognize shading issues and their impact on the PV system. A little shading can result in serious consequences when the designer does not understand the impact to cells, panels, and strings.

Airflow

You must understand the effects of temperature and airflow. Airflow is critical to PV system performance. You must provide at least a 4-inch air gap between the roof surface and the clear airflow portion of the panel frame, where temperatures can be about 90 degrees Fahrenheit or greater.

- Standard test conditions (STCs) test panels only under specific and narrow conditions. It is important to figure out how panels will perform under real site temperatures. To create a well-functioning PV system, a designer must understand the following points about airflow. Designs that promote laminar flow reduce temperatures better than designs promoting turbulent flow. Turbulent airflow creates array hot spots.
- Airflow on arrays continually changes as the wind direction changes. Do not assume that bottom-to-top airflow conditions are common; they are not. This is why array hot spots are always a function of cell temperature and not ambient temperature.
- Sealed designs mounted flush reduce or eliminate airflow. This significantly reduces energy production and shortens the life of the panels. This is especially true for roof tile systems or flush-mount kits with short standoffs or mounting brackets.
- Battens for roof tiles do not provide sufficient airflow in moderate to high-temperature environments. This includes most of North America. If you have to deal with battens, you should further derate, or discount, system performance. This will affect string sizing and inverter selection.
- The cost of providing effective airflow is minimal compared to the substantial reduction in performance. Optimum airflow is important for the life and performance of the PV system.
- Technologies that are not as sensitive to spectral responses in high temperatures will have a reduced usable life.
- Always request the manufacturer's temperature test data. This is important for locations where temperatures will be above 90 degrees Fahrenheit for at least 30 days per year.
- Avoid using micro inverters affected by low airflow in locations where temperatures will be above 90 degrees for at least 30 days per year.
- Systems designed to trap air so that it can be drawn off to heat air or water will have airflow issues. This will reduce system performance, cause hot spots on the array, and shorten panel life.
- Consider canopies and parking structures that use panels with the backside open to airflow for better performance.
- Always provide sufficient airflow.

Wind

Wind can assist in cooling panels and raising performance. Wind can also negatively affect a system by creating uplift and torque. PV system design must account for the extreme wind conditions, not merely average ones, where the system will be installed and operating.

When designing a PV system, the designer should keep the following wind-related points in mind:

- You should develop a complete knowledge of the local wind conditions. This should include issues such as tornadoes, hurricanes, and micro bursts of rain.
- You must design for the highest historic wind level for the site and region. Winds in excess of 120 miles per hour may result in debris damage to panels. Panels and mounting hardware exposed to winds from 120 to 150 miles per hour will probably be destroyed by flying debris. Owners should consider a strategy for removing panels if in the path of a hurricane. Most mounting systems allow for quick removal.
- Wind seldom flows in a continuous pattern. This inconsistency will create hot spots on the strings.
- Always consider design strategies that reduce the negative effects of hot spots.
- Pay attention to roof corners and other areas vulnerable to the Bernoulli effect. Pay particular attention to this for ballasted systems.
- The most unexpected impact from wind may be from another building. Local topography can adversely affect wind patterns.

Space Available

Conducting an accurate space study is as important as conducting the load profile, sizing the system, and determining the budget. Besides panels, inverters, and other physical components of the PV system, understanding the spatial and string sizing issues, obstructions, impediments, and existing or future shading challenges will define the true amount of space available for the system array or arrays.

To make the most of the space available for the PV system, it is important to understand the following:

- Space permitted for the PV system will directly affect the performance of the PV system's panels and inverters.
- High-performance systems have a higher energy-collection density and deliver more energy in a smaller area.
- Space available includes physically usable space that will not be shaded or negatively affect the PV system during its life. The system needs enough

space to meet installation codes and safety requirements. Install the PV system and components where they are not at risk for damage.

- Space considerations include future additional creative PV applications such as building-integrated PV (BIPV), parking, and shade created by structures, trees, mountains, antennas, or other shade-producing objects.
- Understand the importance of not obstructing airflow or access to HVAC equipment, chimneys, vents, skylights, walkways, or other priority equipment.
- Understand how to avoid possible damage to the PV system from snow, water, and organic debris. Understand how to avoid any secondary damage from the design's spatial preferences.

Structural Challenges

Collecting accurate structural measurement data will help prevent expensive mistakes and possibly dangerous situations with the PV system. Consider the following when collecting structural data for the PV system:

- Structural considerations include the system's mounting hardware, the surface where mounting hardware is attached, wind conditions (especially with respect to uplift), temperature, and service access to the system.
- If structural information is not available, consult with a structural engineer to determine what the structure can and cannot support.
- You must possess an in-depth understanding of historical climate conditions because they affect the physical system.

Orientation

For most fixed installations, panel orientation is considered optimal with a true south–facing direction at a tilt equal to the location's latitude. System designers seldom have the luxury to meet those conditions, however.

Consider aesthetics when orientating the PV system. It is important to predict system performance accurately. Many project sites do not allow for a perfect-south orientation. It is essential to understand the impact of being off true south.

The penalties become greater as the panels move away from true south. Listed here are issues related to true-south orientation:

- It is critical to be able to properly use a compass and accurately adjust for magnetic declination—the difference in the number of degrees between true north and south.
- In environments where fog or shading is an issue in the morning or afternoon, the design and installation must consider those facts.

- Flush mounting on a roof may not favor optimal performance. Understanding the impact of flush mounting is critical to effective design and installation.
- Aesthetics may affect array slope or other orientation considerations, and must be seriously considered and addressed in any analysis.
- You must understand and communicate to the client all the impacts and consequences of orientation.
- Avoid compromises when possible. These include mixed-string orientation and slope on an inverter or low-production orientation. Always avoid multiple orientations for strings on the same inverter.
- You must master the impacts of orientation.

Mounting

When mounting equipment, you must consider all the related structural and environmental issues that exist at the site. Appropriate design results in a mount that will outlive the useful life of the system.

Understanding the following points will assist the PV designer in creating a securely mounted PV system:

- Properly torque all mechanical connections during installation and after burn-in; then check annually.

Two men support solar panels as they install them on the roof of a home. This 2007 Solar Decathlon competition home includes residential photovoltaic (PV) modules from SunPower Corporation of San Jose, Calif. and a solar collector from koTech of Budapest, Hungary.
Courtesy of DOE/NREL. Credit: Jim Tetro

- Additional wind flow at the corners of buildings, especially for ballasted systems, will require reinforcements and additional anchoring at these points.
- Be 100 percent certain that all penetrations or connections to a roof, building, or other structure are performed so in a structurally complete and sound manner to maintain waterproofing.
- Retorque connections within one year of system commissioning to ensure they are structurally sound and secure.
- Do not install mounting hardware such that uplift and compression over time on a non-perpendicular member will result in failure. That failure will result in damage to the mount, the panels, and the roof.
- All mounting on a roof surface that is under warranty requires the roofer to agree on a mounting method such that the roofer can complete the dry-in and continue the roof warranty. The solar integrator has roof responsibility for up to 10 years after the installation, even if the roofer covers the warranty on his or her work and there are PV system additions or changes to the roof.

Maintenance to the Structure on Which the System Is Mounted

Professional PV integrators are responsible for pricing the PV system. When pricing a system, you must include performance characteristics of the PV system. That is, you must not price the system before you have a clear understanding of the load size, equipment types, array size, and customer needs.

Consider the impact that the system will have on the building and structure. Also, think about all other objects around the system installation.

With regard to maintaining the structure on which the system is mounted, keep the following points in mind:

- Design all structural connections for longevity and full system life.
- The system designer should consider maintenance and/or damage to the existing roof or structure on which the system will be mounted. The impact that the design or installation will have on the total cost of the PV system, as well as the system's operating and maintenance costs should also be considered.
- Consider the structure to which the PV system will be attached during the design and installation phases. You must make accommodations for access for servicing mechanical equipment, making roof repairs.
- When designing a PV system, also allow for safe access to non–PV maintenance items.
- You must design and install the PV system in such a manner that painting, repairs, and regular maintenance such as cleaning (see photo) are possible

A PerfectPower employee cleaning modules.
Courtesy of PerfectPower, Inc.

without having to remove a substantial number of panels or other equipment.

Aesthetics

Primary design and installation include aesthetics concerns. You must properly address those concerns. Aesthetics is about value. Value is both financial and emotional.

Understanding the following points will help the PV designer and the installation team create and install an aesthetically pleasing PV system:

- Aesthetics are a prime design and installation issue. Aesthetics are a factor in all the system components and any impact they may have on the surrounding area.
- Designers must design the PV system to fit properly in the site's environment (see photo). Fitting the PV system to the site will make it look visually balanced. Visual balance will help meet the aesthetic requirements of the client. Visually pleasing PV systems help property values.
- Equipment must fit in with its surroundings whenever possible.
- You must install panels, inverters, boxes, switches, conduit, and all other system components properly. Components should be installed plumb and level, and placed in locations with the least visual impact. If it looks sloppy outside, it probably is on the inside, too.
- You can paint conduit boxes and other components to match the building's or structure's color or texture. Do not paint safety labels or signs.
- When possible, you should match the color of sealants to the surface to which they are applied. You should apply sealants neatly. Too much sealant is seldom appropriate!
- Housekeeping at the installation site is important. You must remove all trash, waste, cuttings, scrap, and other debris daily and at the end of the job. Leave the job site looking better than when you arrived.
- The job site must look professional, clean, and organized during the workday and after the workers leave the job site for the day.

Solar shade mounted by PerfectPower in Phoenix, Ariz.
Courtesy of PerfectPower, Inc.

System Life/System Service Life

As you read earlier, there are three grades of PV systems: subprime, moderate, and high performance. To assist clients in making the best choice in terms of which grade to purchase, it is important that the designer understands the differences and is able to communicate those differences to the client.

In making their purchase decision, clients must consider price and other factors such as system life, maintenance costs, technical upgrades, and possible future expansions. When you educate clients on these issues, they will often choose a moderate or high-performance system. These two types of systems are less expensive to install, operate, and maintain than a subprime PV system.

To assist the client in making better choices, the PV designer should understand the following:

- Designing a PV system that can last 20 years is very different from designing a 40-year system.
- Designing a 40-year system requires a higher, more reliable grade of engineering skill, installation skill, and equipment.
- Higher-performance systems result in better life cycle benefits due to lower operating and maintenance costs, fewer service calls, and a more reliable, more robust system.

- Most subprime systems do not pay for themselves. The unforeseen costs of a system with a short life include higher and more frequent maintenance costs, as well as early replacement. This results in unhappy clients.
- Serviceability includes any costs to the client due to design or installation flaws.

Warranty

Clients expect warranties when purchasing expensive items like a PV system. Warranties can come from the PV system company, the manufacturer, or the installer. Sometimes, local laws (jurisdictional) will require warranties. A true valid warranty reflects the warrantor's confidence in their product or service. You should base warranties on experience, knowledge, and skill.

Following are various types of warranties for PV systems:

- Manufacturers of solar panels tend to give product warranties that are good for one to 10 years on manufacturer workmanship and defects, and five to 25 years on performance.
- Inverter manufacturers usually give product warranties for five or 10 years. Many provide extended warranties for an increased price.
- Jurisdictional warranty requirements tend to be weak. They do not reflect the potential of the system integrator or the PV system's ability to produce energy.
- A system-performance warranty specifies that the PV system will have a certain amount of output for a specific amount of time. These require close monitoring and are uncommon. They will become more common, however, as clients find their systems do not meet the numbers projected by the integrator.

Utility and Energy Production Rate Structure

Utility rates, and the offsetting system electrical energy production, drive PV economics. Integrate utility rates, tariffs, and policies into PV design. Tariffs and rates determine the size of the PV system.

Size and performance determine how fast the system pays for itself. Energy offset from the electrical grid by the PV system will change the tariff structure and rates for the energy system. The utility company will reclassify the tariffs.

You must understand how the following elements affect the PV system:

- The utility rate structure, the PV system's effect on those tariff rates, and the effect of the rates on the system.
- The tariff rates and any potential changes in those rates, especially over time.

- The time-of-day rates and their impacts on the PV system output.
- The utility company's record for pricing, support of renewables, prospects for long-term viability, responsiveness to new technology, and ability to accept clients with grid-tied PV systems.
- Impact of inflation on utility rates and the reliability of the PV system being installed (this can be a selling point)

Here are a few tips to keep in mind:

- Time PV system energy production to occur during the day when energy tariffs are highest.
- Beware of limited-period tariff programs.
- Determine whether the client has a net metering program to work with. If so, take advantage of those opportunities and compromises.
- Study and understand the PV system's impact on demand rates and provide the best analysis for the system's impact on those rates.
- Understand the fee structure that underlies the utility company's cost for kilowatt-hours.
- Understand how the shortcomings in the utility company's structures can negatively affect the client or system economics (another possible selling point).

CHAPTER 1 SUMMARY

This chapter gives pointers on how to sell a PV system and an explanation of individual positions in the PV profession and their responsibilities. This chapter provides an overview of PV system criteria that will continue in Chapter 2.

KEY CONCEPTS AND TERMS

Array

Inverter

Module

Panel

Performance test conditions (PTC)

CHAPTER 1 ASSESSMENT

Overview of Advanced Photovoltaic (PV) System Design and Design Criteria

1. The sale of the PV system to the client is the sole responsibility of the sales professional.
   - ❑ A. True
   - ❑ B. False

2. The location of the PV system affects which two aspects of the PV system? (*Select two.*)
   - ❑ A. Design
   - ❑ B. Local laws
   - ❑ C. Solar energy used
   - ❑ D. Performance

3. Warranties are the responsibility of which of the following? (*Select two.*)
   - ❑ A. Local legal system
   - ❑ B. PV companies/installers
   - ❑ C. Component manufacturers
   - ❑ D. The client

4. Altitude can result in which two of the following? (*Select two.*)
   - ❑ A. Faster system burnout
   - ❑ B. Quick cooling of the module at night
   - ❑ C. More difficult installation of the system
   - ❑ D. Faster heating of the module during the day

5. Which standards do not apply to the PV industry?
   - ❑ A. Industry standards
   - ❑ B. Federal standards
   - ❑ C. State standards
   - ❑ D. Local standards
   - ❑ E. None of the above

6. _______________ has an impact on all local environmental factors of the PV system.
   - ❏ **A.** Altitude
   - ❏ **B.** Climate
   - ❏ **C.** Ambient temperature
   - ❏ **D.** Cell temperature

7. The relationship between the sun and Earth is constantly changing.
   - ❏ **A.** True
   - ❏ **B.** False

8. Which system is the higher-quality, higher-output PV system to install?
   - ❏ **A.** Subprime system
   - ❏ **B.** Moderate performance system
   - ❏ **C.** High-performance system
   - ❏ **D.** 40-year–life system

9. Which two of the following are critical to a business's success? (*Select two.*)
   - ❏ **A.** Shorting the client on quality of service
   - ❏ **B.** Ethical treatment of clients and suppliers
   - ❏ **C.** Providing excellent client service at a fair price
   - ❏ **D.** Shorting the suppliers of the components of the PV system

10. Which of the following data are important to collect regarding climate?
    - ❏ **A.** Local temperature
    - ❏ **B.** Cloud cover
    - ❏ **C.** All microclimate data
    - ❏ **D.** Days under freezing
    - ❏ **E.** All of the above

Evaluation and Design Criteria Part II

THIS CHAPTER REVIEWS THE 36 criteria introduced in Chapter 1 and expands on the information by focusing on how to assess and evaluate each of the criteria items.

Topics & Concepts

This chapter covers the following topics and concepts:

- Monitoring
- Operations and system maintenance
- Cleaning (panels, filters, and equipment cavities)
- System performance modeling
- Quality of installation
- Testing schedule and procedures
- Quality assurance (continual reevaluation and review)
- Service entrance section (SES)
- National and local electrical codes
- Panels
- TÜV, UL panel certification, and other certifications
- Strings, string sizing, and string placement
- Inverters
- Balance of system
- Wire run and sizing, wire-run locations, and protection
- Equipment burn-in response

Goals

After completing this chapter, students are expected to be able to:

- Describe the basic principles, methodologies, and strategies for sizing photovoltaic systems to code and for performance
- Evaluate PV systems and design using key quality performance criteria

Monitoring

Monitoring helps you maintain PV system value and performance and determine system output. The hardware used for monitoring is inexpensive or free.

With regard to monitoring, the integrator should ask the following:

- What is it worth to know whether there is a problem with a system before the client does?
- What is the value of being able to look at the same data as the customer to see whether there is a problem?
- How much will monitoring save in service calls?
- What kind of information does monitoring provide?
- Can the integrator and the client both access the information?
- Can the integrator link all clients under one account to view their data quickly and readily download the data for analysis?
- What kinds of data export options are there?
- Is there software that will review and compare the data to flag a problem?
- How is the system designed for the transfer of data—wired, wireless, or both?
- How long do Internet-based monitoring systems store data, and how far back can it be accessed?
- How much will Internet access for system monitoring cost?
- If monitoring is free, how long will it be free before a fee is charged?

Operations and System Maintenance

Operations and system maintenance includes all the functions that take place after commissioning. Lowering operations and maintenance costs is part of the art of systems design. It goes hand in hand with a top-quality installation. When done properly and regularly, operations and system maintenance will recoup more than their cost in additional energy produced. Operations and system maintenance will also extend the life of the system, providing additional savings.

Serviceability of Equipment

The correct choice and placement of equipment are important to the serviceability of that equipment. Consider an example in which a combiner box is installed on a tile roof. In such situations the integrator assumes he or she will never have to touch it again. However, if the integrator ever has to service that combiner box, the team will have to walk on the roof to get to it. This may result in broken tiles, which represent an unnecessary expense. If you do not properly install the combiner box to allow for easy access for testing, string checking, or modifications, you could see an added expense in the form of additional roof work. One solution is to install the combiner box as close to the strings or array as possible, while placing it so that testing, torquing, and any modifications can take place without risking the roof tiles.

As a second example, consider what might happen if you install an inverter in a dark, narrow walkway or hallway. The darkness of the location makes it difficult to see the telltale signs of a DC short. You should install inverters where the service technician can clearly see them, as well as see all disconnects and connections.

Sometimes, achieving perfect product placement is difficult. Even so, when designing and installing a PV system, keep in mind that technicians will need to access the system's components.

Life of Equipment

A moderate-performance PV system can easily last 20 years with moderate levels of maintenance. But high-performance systems and quality work will provide so much more value. Here are a few points to keep in mind about equipment life:

- Be aware that inverters are a weak point in PV systems. But the right inverter will last 15 to 20 years, unless there is a defect in the design, installation, or product. The integrator should understand how the inverter lives and dies.
- Panels are the most expensive portion of the system. But they can easily last 40 years based on performance design when installed properly.
- Performance of the PV system will drop over time. However, if the system is properly operated and maintained, defective panels and other components will be replaced under warranty, allowing the degradation to be less than is often assumed.
- A 30-year-old high-performance system, operating at 75 percent capability, will perform as well as a mediocre new system. There is incredible value in

choosing the best quality products, design, and installation from the outset and throughout the life of the system.

Required Maintenance

Performing the required maintenance results in the following benefits:

- It extends the life of equipment.
- It increases the lifetime kilowatt production of systems.
- It allows for the replacement of defective and underperforming equipment under warranty.
- It allows for the replacement of defective and underperforming equipment that is not covered by warranty early in the deterioration process, thereby keeping output higher.
- It continues a fine-tune process that keeps production up.
- It enables the integrator to gain experience and knowledge and to sharpen troubleshooting skills. This assists the integrator in developing new techniques and strategies for problem solving and gives the organization an image for thoroughness and enhanced capabilities.

Cleaning (Panels, Filters, and Equipment Cavities)

Cleaning PV system equipment can result in a 5 percent (or more) improvement on system performance. For one thing, keeping equipment clean will reduce heat buildup on components. Heat buildup shortens equipment life. Proper maintenance—of which cleaning is an important part—always pays for itself and can even save money.

Cleaning involves removing what settles on, covers, or invades the equipment. This is often a mix of dust, dirt, hydrocarbons, and bird droppings. Other soils include leftovers from other maintenance tasks like roofing materials and sealants, tools, or HVAC components. Note that rain and snow will not keep panels clean.

Of course, environmental conditions vary by location. For this reason, the cleaning section of any maintenance plan must be specific to the system site location. Note that a maintenance plan requires consistent record keeping.

A PerfectPower employee cleaning panels in Phoenix, Ariz. *Courtesy of PerfectPower, Inc.*

Panels and General Equipment Cleaning and Inspection

You must perform a site evaluation to determine how much cleaning, and of what type, will be required.

You should consider the following points when preparing for a site evaluation:

> You should train maintenance personnel to clean and test systems, and to identify emerging or existing problems. The sooner problems are fixed, the more positive the economic effect on the PV system.

- Is the site urban or rural?
- Is the site subject to dry, dusty conditions or mostly wet conditions?
- Is the site near a freeway, airport, industrial area, construction area, or other area with airborne hydrocarbon or other chemical products or residues that can stick to the panels?
- Will the quality of local water allow for a quick, clean rinse?
- Does cleaning require a solvent? If so, how often?
- If a solvent is required, which one will clean the panels without leaving a compromising residue, which might make the situation worse?
- When is high-pressure water sufficient, and when does a brush need to be used?
- Are there exceptions to the cleaning schedule?

Filters

Dust and chemicals have a negative impact on the inverter. Many commercial and industrial inverters use a variety of filters to remove potentially damaging dust and airborne adhesive chemical compounds. Some residential inverters and charge controllers also provide filtering, which safeguards equipment.

Following is a list of things to consider with respect to cleaning filters:

- You should review the manufacturer's recommendations for information about installing, cleaning, and replacing filters.
- You should review the condition of the filter(s) monthly to determine the level of filter soiling. This will help to establish the cleaning and replacement cycle. It is better to replace or clean a filter slightly earlier in a regular cleaning and replacement cycle than later.
- You should track filter condition and soiling levels with accurate records.
- You should consider the level of difficulty in cleaning or replacing filters.

Equipment Cavities

Components of PV systems have cavities that animals, insects, and reptiles can access. These critters can create a short that can damage equipment. Use caution when opening inverters, combiners, switches, and the like. Make sure there are no insects, reptiles, animal nests, or dead animals in the cavity. If there are, remove them very carefully. If necessary, contact an exterminator. Remove debris like leftover dirt, dust, tools, and wire or insulation clippings from the cavities.

System Performance Modeling

System performance modeling is a tool that provides accurate options for system design. It is imperative that you become adept at using modeling software. You must grasp its strengths and weaknesses. You must understand its algorithms, input and output, and how it processes the data supplied. You must learn how to make adjustments to get the highest level of accuracy. It is also critical that you understand the philosophy under which the software was developed. Understanding the software helps you to understand the data output.

Following are some points pertaining to modeling software:

- Accurate modeling requires education and training on design models, which are complex. The better the designer and the team understand the model, and how to provide the most accurate input data, the higher the modeling accuracy with respect to PV system performance. You must use the best, most accurate models available to you.
- You must enter data properly, and the data must reflect real conditions.
- Modeling is only as accurate as the data entered into the program. Input does not necessarily reflect all the details needed to produce the greatest value.
- Use an accurate model that will work for the environment. Back up system design with real monitoring data. This will validate the modeling.
- Modeling requires a commitment to design, process, equipment selection, quality installation, and more.
- Two designers or installers can use the same equipment and design to arrive at different results. Consistency and interpretation are important.
- Take time to understand the local environment in which the PV system will operate. Local meteorological data is helpful, but site data is required for accuracy and clear, consistent results. It is a design challenge to interpolate between the two, often with limited information.
- Input accurate data based on local environmental conditions. Entering irradiance, weather, and microclimate information is extremely important in terms of accuracy.
- Relate accurate data to the required input.
- Understand the impact of each input value to true system performance. This will enable the PV designer to make incremental changes in final performance.
- Review pre-entered data; then review it again for accuracy.

- Derating factors are developed at STC. Know how to adjust for real conditions, especially for derating coefficients. Reviewing data accuracy is important for revealing errors. Always have at least one very knowledgeable individual review the data for accuracy.
- Double-check installation to ensure all bolts are tightened, connections are made properly, etc. This will help ensure everything is completed and the system will be ready for use.

Quality of Installation

A quality PV system installation includes the following:

- A combination of stringent requirements and standards
- Training and education
- A complete understanding of the installation team's skill

An installer will complete between 1,200 and 2,000 hours of training. This includes classroom training, lab training, and direct supervised instruction. You should combine training and quality assurance. This eliminates problems during the initial installation phase.

Following are some considerations with regard to installation procedures:

- Having more trained eyes on a project will reveal seemingly insignificant issues that collectively affect the system's operation and performance.
- It is a good practice to reward installers for a combination of education, system performance, and customer satisfaction (evidenced by a lack of callbacks).
- Consider offering profit-sharing benefits on a per-project basis for installers. Determine early on what mistakes and callbacks cost the company, and share a portion of your earnings with exceptional installers whose work mitigates those costs.
- The greater the installer's knowledge, the stronger the team will be. Teams that reinforce each other throughout the project create a better PV system.
- It is extremely difficult for an untrained or semi-trained laborer to grasp the importance of the details required to install a good-quality system. Much of the work may be repetitive. Installers who do not understand the importance of the various details will hurt the final product.

Testing Schedule and Procedures

Adopt a strong quality-assurance process and program. This will result in a reduction in the number of system defects. It will also reduce performance losses, and result in fewer service calls. Quality assurance and testing involve a long-term commitment.

It is crucial that you implement testing procedures. Testing finds defects in your product and in your installation routine. Reducing defects by 5 percent can add a year's worth of kilowatt-hours of output. Testing also helps to fine-tune the system, which in turn increases output and lowers operating and maintenance costs.

As part of the testing process, you should perform a visual inspection. Ensure that conductors are properly installed and check torquing connections.

Pre-Commission Testing

You should test the PV system before commissioning the system. Then check the system again before local code enforcement inspections. These checks should include (but are not limited to) the following:

- Test each string for voltage and current. Record the information.
- Compare each string. Check any string that performs outside a 2 percent variation for connectivity. Also, check to see if there is a low-performing panel in the string.
- Inspect every wire, conductor, and cable path.
- Confirm labeling.
- Confirm that all the connections have been properly torqued.
- Check for pinched conductors.
- Check for proper string connections.
- Check all conduit and conduit connections.
- Record the results using a responsible party that guarantees each test.
- Complete and confirm all continuity testing and record appropriate voltages and current for each string, for each combiner set, and for home runs.
- Use an infrared camera to look for hot spots, bad cells, or irregularities in the array. These items will show themselves clearly in the infrared wavelengths.

Post-Commission Testing

PV companies need to stay involved with the systems they install after commissioning. Activity should continue for the life of the system.

Burn-In Period Testing

The burn-in period spans the 90 to 100 days after system installation. This enables the system to go through several thermal cycles. During this phase, mechanical and electrical connections will set or loosen. You should inspect all connections, visually and physically. Retighten and torque loose connections. Look for any potential electrical shorts, gaps, damaged equipment, and crossed cables. Thermally inspect PV panels, strings, and arrays.

Monthly Monitoring

Monitor all PV systems weekly to monthly. This will guarantee the integrity of the system. It will also draw attention to performance issues.

Annual Testing

Annual testing can be visual or more like the burn-in inspection and testing. The larger the system, the more cost-effective it is to perform a more thorough test regimen. At the minimum, the customer can do a complete visual inspection.

Quality Assurance (Continual Reevaluation and Review)

One of the secrets to providing professional-quality, high-performance PV systems is mastering the process of quality assurance (QA). Quality assurance eliminates as many defects and challenges as possible from design and installation. Providing QA will increase the client's confidence in the PV team. As such, the PV team should take QA seriously. Performing QA on the PV system will give the PV team a competitive edge in the marketplace. Of course, there is a cost for implementing an effective QA program. In the long run, however, QA saves money. Planned service visits are less costly than unscheduled ones.

The best approach is to develop a continual quality-assurance process as the project evolves. All participants should continually review the actions, activities, steps, and procedures. That way, if they miss a step or fail to properly use a technique, the issue will be addressed before that set of steps is complete.

If you reward the team for perfecting their skills and installations, fewer irregularities will occur. Under the watchful eyes of an advanced journeyman or master PV professional, all quality-assurance steps and procedures must be taught.

Following are steps you can take toward continual reevaluation and review:

- Ensure that each professional PV installer at each level is familiar with the standards of continual quality assurance.
- Offer installers an incentive to perfect their professional skills. Incentives can be in the form of professional recognition, advancement, bonuses, profit-sharing when jobs are completed ahead of schedule or under budget, tickets to a sporting or music event, or any other incentive that appeals to the team.
- Ensure that a highly skilled advanced journeyman or master PV professional (also known as the project lead, supervisor, superintendent, or project

manager) leads the team. He or she must oversee the project while becoming the focal point for minor system corrections and adjustments to maintain quality.

- Review each system segment upon initial completion and at the end of the installation, prior to startup.
- Ensure that the lead journeyman or master PV professional checks, records, and signs off on each section of the review. You should keep these records with the permanent project files.
- Test and balance each string, taking care to record all data. You should address any anomalies immediately and record all changes.
- Check each combiner box for consistency, irregularities, and defects. Be sure to record all the data you collect.
- Check voltage and current passing through the combiner box for a baseline. Be sure to record all the data you collect.
- Check all connections for torque foot-pounds. Be sure to record all the data you collect.
- Visually check all conductors, and test them for continuity, termination, and labeling. Confirm their size and record data prior to startup. At this point, you should identify and correct any improper connections, torquing, or labeling, and any imbalanced, mixed, or damaged strings.
- Visually and manually inspect all additional connections. Record the datTest arrays with an infrared camera. Look for hot spots, bad cells, or any other irregularity that appears in the infrared frequency range but is not visible to the naked eye. Record and save those images for comparison at burn-in and for each additional operations and maintenance (O&M) review in the future. This data may be helpful in a warranty case or other type of claim.
- You should perform a system review 30 to 60 days before the end of the warranty for the inverter, panels, and any other covered components. This enables you to replace faulty or sub-performing components before the warranty ends. You might provide this to customers as a service option, for a fixed fee. This service is valuable, especially when there are items that can be replaced under warranty.

Service Entrance Section (SES)

During the design and installation phases, substantial consideration for a service entrance section (SES) may be required. Although you can sometimes use an SES without modification, modifications or replacements are sometimes necessary. Supply-side taps are sometimes required.

Understanding the following points will help you design SES into the PV system:

- Understand the SES. This includes availability of parts such as breakers and fuses. Understand quirks in the system.
- Understand the local utilities requirements. Some will be negotiable.
- Understand how the local utility company and local jurisdiction address SES and other issues. Sometimes, the utility company and local code enforcement interpret code differently from each other.
- When running into serious challenges, ask the jurisdictional and utility specialists for assistance. Inspectors expect questions from designers and installers.

National and Local Electrical Codes

PV designers and installers have to follow the National Electrical Code (NEC) and local electrical codes. These codes are legal requirements, and outline the minimum standards for safety. Exceeding these codes will increase the system's safety, performance, longevity, and reliability. Mastering the electrical codes will save the PV designer and installer time and money.

To master the various electrical codes and establish positive relationships with inspectors, you'll need to understand the following:

- National and local electrical codes
- The meaning of codes
- The codes' history
- The reasons specific codes exist

Cutting corners on existing codes places lives and property in danger. Just because you don't understand the reason for the code does not mean that it has no purpose.

Work with local code officials, including plan reviewers and inspectors. Establish a working relationship early in the process. Inspectors will tell you what they are looking for and what is acceptable. This will save you time and money. You should also consult with plan reviewers and inspectors at the beginning of the job. You must understand their requirements to succeed.

Panels

There is a wide selection of PV panel technologies on the market. It is important to select the one that best fits the project's environment and will deliver the results the client wants. All panels and technologies have their own architecture, peculiarities,

PV panels installed by PerfectPower in Phoenix, Ariz.
Courtesy of PerfectPower, Inc.

strengths, and weaknesses. It is up to the PV designer to understand how the technologies work and to incorporate the right panels and technologies to achieve the desired results.

All PV panels are tested and certified at STC. The manufacturers in their equipment brochures publish test results. STC test panels perform under standard test condition, not at actual temperatures in real-life environments. The PV designer must incorporate both STC data and real-life data to select the best panels for the PV system.

Be aware that American design software uses air temperature to size strings and design systems. This is in contrast to software in Europe and other locales, which uses cell temperature. In PV design, cell temperature gives more accurate design feedback and much better results.

Temperature and spectral-response issues are critical to performance. You must consider them in the evaluation and design phases over all the environmental conditions.

Selection

There are numerous manufacturers of PV panels. When selecting PV panels, the PV designer should consider the manufacturer's product development, manufacturing experience, and reliability. The designer must understand each product's strengths and weaknesses. This will enable the designer to differentiate between manufacturers. This is necessary for the designer to be able to select a product to use in the PV system.

Installation

Installation techniques are critical to the success of the PV system. Following are guidelines that will assist installers in installing a high-quality PV system:

- Protecting panel cables for the life of the system (usually 40 years) is essential to the performance of the PV system. Protecting the cables eliminates performance issues. It is also an important safety issue. Proper protection is the result of design and installation techniques.
- Properly connect and secure cables with an appropriate clip. An appropriate clip will not deteriorate or break. You must protect connector cables from environmental conditions like the sun, heat, cold, thermal expansion and contraction, wind, rain, snow, ice, and animals.

- Protect cable, wire, and any insulated conductors from the sun. Never expose cable, wire, or conductors to direct sunlight.
- Inspect the panel's mechanical connections after burn-in and during the second, third, fourth, and fifth years of operation. Secure and repair any loose connections.
- All panels and equipment must be installed level and plumb. This helps make the PV system aesthetically pleasing. In addition, give a balanced and even presentation. One of the few exceptions will be special architectural pieces and PV art.
- When installing cables, create clear visual paths so that cables can be visually inspected. It is important to ensure that cables are not pinched, and that they do not create a dangerous short or a ground fault situation.
- Protect conduit from high temperatures. Follow NEC temperature guidelines and calculations for placing conduit above a surface and provide additional air space. Provide extra safety precautions in locations where high temperatures will occur. The NEC is the minimum requirement.
- Install all mounting hardware, including struts for racking, on the same slope, or angle, and on the same plane and position. The relationship between visual propriety and structural value of the mounting hardware will make the arrays more visually appealing.
- Confirm that the plans are correct. Verify that each array and its components are not shaded, and will never be shaded.
- Inspect to make sure the panel is properly installed. This means making sure the panels are not twisted, torqued, or improperly attached.
- Clean panels after installation is complete. Cleaning will remove oils or other substances that may attract dust and dirt, thereby affecting system performance.
- Carefully check and record array-voltage and current readings. Compare voltages to other arrays on the same inverter. Keep variations at a minimum. This will maximize system performance.
- Visually inspect cells, bus connectors, and panel components for discolorations and blemishes. Blemishes can negatively affect performance.
- Take infrared pictures of the panels and arrays. Confirm there are no hot spots. Use those infrared panel images for comparison later as needed.
- Integrators need to have clear communication with the PV designer and installers to uncover inconsistencies or problems during the panel-installation process.
- Panel installation must be consistent with a mounting system that will properly secure the panels over the life of the system.

TÜV, UL Panel Certification, and Other Certifications

American standards and certifications do not include reliability standards for equipment, design, or installation of PV systems. Currently, there are more than 5,000 panel manufacturers worldwide. The quality of products produced by these manufacturers can differ greatly. It is important for PV professionals to research the products they want to use to create a PV system.

Following are points to consider when choosing materials:

- Understand the required panel certifications, what they mean, how they are tested, and what the conditions for certification are.

  - Always choose panels with the most stringent certifications. For example, select panels that include certification from both UL and TÜV, the German Technical Inspection Association. TÜV has a reliability component not found in UL, even though UL is required in the USA.
  - Certification testing is not the same as performance standards. It is important to review the industry information on the products being considered.
  - Spend time at a testing lab to see how testing is conducted. Then read the testing reports from the lab. This will give you insight into how to recognize quality panels by comparing manufacturing technologies. Compare product technologies and their failure rates by class.
- It is always better to ask what the problems are with a group of products, and not just rely on the marketing materials.

To increase the quality of the PV industry in the United States, encourage associates and competitors to support the adoption of clear, stringent, and effective design and installation standards, especially with regard to reliability. This will help expand PV by giving potential clients more confidence in the industry.

Strings, String Sizing, and String Placement

String sizing is based on the following:

- System design
- Installation goals and objectives
- All environmental conditions
- The technology selected for panels and inverters
- The system's maximum power point tracking (MPPT) capabilities

The **maximum power point tracking (MPPT)** function electronically forces the inverter to accept power from the PV panels and strings at their greatest power production point on the **I–V curve**. It results in additional power to the system under specific input conditions. **FIGURE 2–1** shows how most PV designers

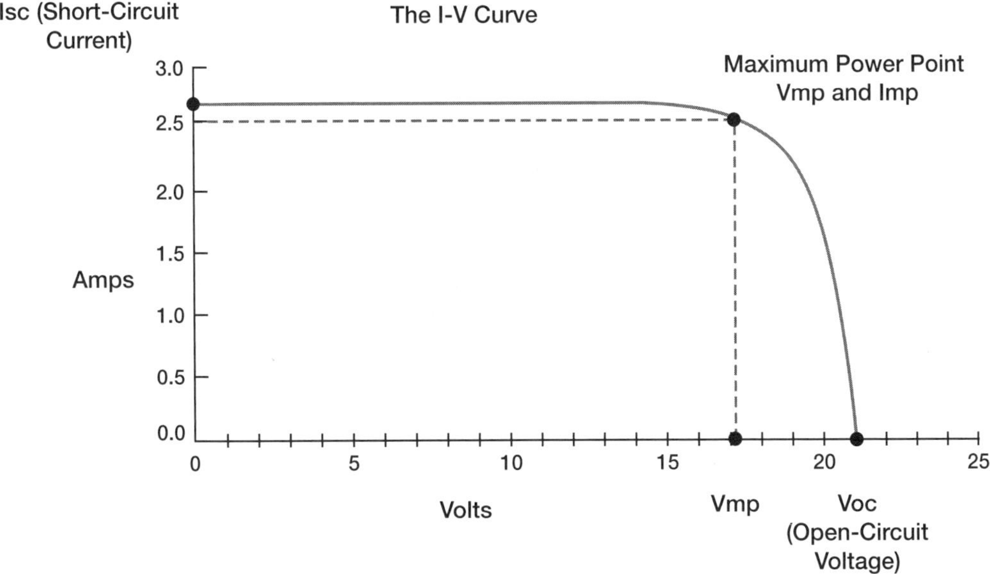

FIGURE 2-1 This figure shows the I-V curve and MPPT.

and integrators see the IV curve in relationship to the MPPT box window. The perception and assumption is that the IV curve is clean and predictably smooth. In this view, the usable energy is inside the box. Understanding what is inside and outside the box is not usually seen as relevant to design. This is because the mental picture is of the I-V curve at STC, not the scattered mix of conditions that exit in the real world.

The MPPT function occurs instantaneously and changes as different products change or retarget the greatest power production point at different time intervals. Conditions in which external conditions change slowly prompt the system to receive the greatest power boost from MPPT. Output is degraded under system conditions where the MPPT does not fall within a specific set of parameters. Energy is wasted when the input DC energy conditions are outside the box. You should maximize conditions and parameters that enable panels and strings to work in harmony with MPPT. This will improve the overall efficiency and power of the system. Higher system performance will be achieved. Waste will be decreased.

Understanding and applying the following points will help you to create an excellent PV system:

- Using accurate input data with a string sizer and code is critical to the PV system's performance.

- You should adjust for real temperatures on panels and arrays. Understand how the efficiency, general specifications, supplied data, cell architecture, spectral response, and all the various real-life conditions will affect how well the PV system functions.
- String-sizing software is used primarily to keep the system within NEC safety standards. This software offers solutions for choosing string sizes and automatically limits string-size choices. The designer can then eliminate the high and low and marginal string sizes that lead to marginal or negative energy production for the system.
- For optimum performance, panels in a string must all be in the same plane and orientation.
- Splitting panels in a string with long wire runs reduces that string's production. It also reduces the energy from the system feeding the inverter.
- Mixing string orientations on an inverter compromises performance, confuses the MPPT software, and results in subprime system performance.
- Understand the influence of the inverter on MPPT. There is a performance characteristic box for MPPT where DC power can pass through to become AC power. Anything outside the box is unusable, and will shorten the life of the inverter.
- Understand when to accept a compromised string design. Be clear with your client why that decision was made and the ramifications it will have on the PV system.
- Double-check the panel cut sheet data for accuracy.
- When panel and string sizing results in power outside the MPPT's range, it's called clipping. Clipping wastes energy and should be avoided.
- Begin system and string sizing by looking at all the possible install locations for panels. Consult with the client to determine what will work and what will not. Give the client a clear choice between moderate or average performance and high performance.
- Understand the frequency between MPPT measurement and adjustment time. The higher the frequency of MPPT adjustments that occurs per minute, the greater the accuracy of the MPPT function. This is a sign of high performance. This is especially critical, as environmental conditions can change rapidly.

Inverters

PV systems use inverters to convert DC to AC. The number of inverter manufacturers grows constantly. When selecting an inverter, it is important for the PV design team to research the manufacturer's claims regarding its product. You should select an inverter that best meets the PV design and load demand.

Keep current with all technical updates on these products. Updates will warn of any technical problems, and have information that will assist in choosing the best inverter for the PV system.

Consider the following when selecting an inverter:

- What are the environmental temperature conditions for operation and warranty?
- What temperature conditions does the manufacturer allow in its warranties?
- With respect to temperature, what is in range and what is out of range for the inverter? What are the consequences of being out of range?
- Does the manufacturer provide additional information on design parameters for extreme conditions? You should request performance data from the manufacturer for temperatures above 70 to 90 degrees Celsius.
- Does the manufacturer always meet the least or lowest certification?
- Does the manufacturer provide responsive and good-quality customer service, training, and tech support for dealers and/or owners?
- Does the manufacturer offer extended warranties?
- Is there sufficient space in the box or cabinet to work, make connections, carry out testing, and detect telltale signs of problems like shorting, hot spots, loose connections, damage, etc.?
- What kind of inverter will work best for your project—with or without a transformer?
- Ask business associates and competitors about problems they have experienced with inverters and other PV products. Learn from their successes and failures.
- Ask other PV professionals about challenges they have faced when installing specific products, and how they have overcome them. Sometimes, installation techniques create problems with inverter operations.
- Understand and communicate the warranties to the client. Be sure to include extended warranties.
- Master all the monitoring equipment. Understand the technology. Know how to install it and connect to the data server. Understand initial costs and 20-year costs.
- Look at the technical details first, make one or two selections, and then shop price. To determine true value, compare products by different manufacturers specification by specification.
- All manufacturers have design and field problems. Find out each manufacturer's history, paying special attention to how issues were resolved. Issues to consider include the following:

  - Was the company open and forthcoming?
  - Did it solve the problems promptly?

- Does it make it easy to find technical reports?
- How responsive is it on warranty issues?
    - Do the staff there want to know about issues? Or do they just discard your feedback?
    - Understand each manufacturer's philosophy about its products and market, and its long-term efforts in technology development.
    - If you find it difficult to get questions answered before you purchase the inverter, be aware that it will be even harder after you purchase and install the inverter.

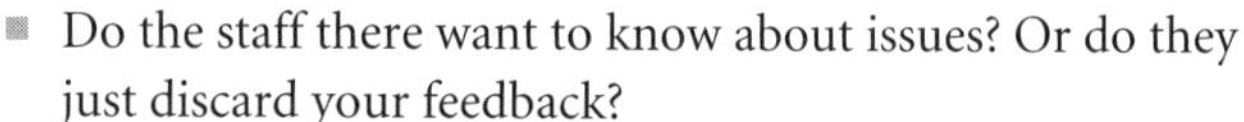

NOTE

Do your best to get to know an inverter's design and applications engineers. Over time, this will enable you to gather information that makes their company the go-to supplier, or eliminates the firm from consideration.

Transformer or Transformerless Inverters

A standard inverter has internal transformers. These transformers are generally heavy, and operate at about 90 to 95 percent efficiency. Under standard inverter test conditions, the transformer resistances are responsible for 1 or 2 percent of inverter loss.

Since 2005, officials have allowed the installation of transformerless PV systems in some parts of the USA. Changes in the NEC opened the door to these inverters, a long-standing technology in Europe and Asia. The first such inverter was Magnetek Inc.'s Aurora PVI, released in 2006. In addition to not having a transformer, transformerless inverters do not need grounding, nor do they produce the additional heat that a transformer typically does. Their manufacturers claim these inverters are up to 98 percent efficient. When transformerless inverters become fully available on the market, they will cost about half what existing transformer products cost.

Differences between transformer inverters (standard inverters) and transformerless inverters include efficiency, weight, and functionality.

Following are some challenges associated with transformerless inverters:

- Transformerless inverters are not allowed in Spain, Italy, and a number of North American locations.
- There has been substantial resistance to transformerless inverters from utility companies.
- There is resistance from regulators and inspectors that will require changes to local codes and standards.
- Plan reviewers and inspectors will need to be educated on the appropriateness, safety, and reliability of transformerless inverters.
- Transformerless inverters have known issues related to galvanic protection.
- All these points create obstacles to significant improvements in the industry, such as lower weights, lower costs, and higher efficiency.

Inverter Loading

Inverter loading can be complicated. It is important not to overload the inverter. Overloading the inverter will shorten its life. Overloading the inverter may also negate the inverter's warranty. Environmental conditions may also adversely affect the inverter. The manufacturer is responsible only for the life and written conditions of the warranty. The solar professional's goal is to provide the lowest cost per kilowatt-hour with the lowest operating and maintenance cost for the life of the system.

Loading factors and conditions beyond the panel specifications include the following:

- High and low temperature on the panels and the inverter
- Irradiance on a seasonal level
- Lensing from passing clouds, which can increase the irradiance input up to 30 percent for short periods of time
- Surfaces that reflect seasonally due to angle of incidence and reflect due to factors such as snow or reflective roofing

Following are several items to consider when determining string sizing and inverter loading:

- Consider string sizing and inverter loading when dealing with high or low temperatures.
- It's safe to size inverters to 80 percent of the array's capacity. Also, you must understand how the array string size affects the inverter's input and output. Remember the MPPT box window.
- Arrays do not feed inverters with DC power in a neat and clean set of power and efficacy curves.
- At low temperatures, the inverter will waste energy though clipping. Overloading will cause clipping year round on most designs. However, if the inverter gets too hot, it has to work harder at 100 percent loading, which will shorten the inverter's life. Loading the inverter at 100 percent reduces energy production and increases maintenance and replacement costs.
- An 8,000-watt DC array can feed a 10,000-watt AC inverter. The inverter will actually receive more DC power in a high-performance system than in a subprime or marginal system.
- There is a variation relationship of more than 20 percent. This is due to environmental conditions such as lensing, extremes in cold temperatures, and albedo. Lensing is the additional irradiance boost of up to 30 percent produced by the edges of passing clouds. Albedo refers to seasonal and temporary boosts in irradiance due to reflection at certain incidence angles from local objects.

The PV system design must account for all these points. In addition, it must account for the fact that a number of losses will occur due to the inverter not being able to process all the DC power. In addition, the PV designer must adjust for limitations in software for determining string and array sizing. The current industry standard is to use string-sizing tools. These standards keep the PV design within NEC code and the inverter load up to 110 to 125 percent. But as mentioned, loading the inverter over 100 percent will shorten the inverter's life and create problems with the PV system.

Following are points to consider when designing the inverter component of the PV system:

- Load the inverter to 80 percent. Loading the inverter to 100 percent shortens inverter life.
- For inverters in moderate to high-performance PV systems operating in 90 degree Fahrenheit ambient temperatures for more than 30 days per year, load only to 80 percent. If the DC wattage is 8,000 DC watts, a 10,000-watt inverter will be necessary. Load the inverter as close to 80 percent as possible. In addition, you should almost never load less than 50 percent for a phased system that is not competed.
- If the inverter operates in moderate temperatures and conditions (between 70 and 85 degrees Fahrenheit), load the inverter to 90 percent.

Inverter Location

With respect to inverter location, here are some points to consider:

- Inverters are sensitive to temperature. For this reason, you should not locate inverters in direct sunlight. Inverters located in the sun will operate in temperatures higher than what they were designed to handle. This may void the inverter's warranty.
- Understand the path of the sun between the equinoxes. In the Northern Hemisphere, the sun rises north of east in the morning and sets north of west in the afternoon. You should make the appropriate modifications for installation in the southern hemisphere. Design accordingly.
- Inverters do not operate at 100 percent efficiency in an air-conditioned space. Efficiency losses create heat. The heat can be used in space-heating conditions. However, you must cool the heat in space-cooling conditions. A 95 percent efficient inverter will pump 5 percent of the DC energy collected into the conditioned space like a little space heater. You will need to remove that energy when in air-conditioning mode. This requires energy and reduces the overall effectiveness of the PV system.
- You should keep large inverters in a semi–air-conditioned space. The space does not need to be set at the same temperature as for people. But keeping

the equipment operating out of extreme temperatures increases its performance and extends its life.

- Use venting to cool inverters. Evaporative cooling air-conditioning can be used. Use passive and thermal mass to control temperatures.
- Inverters require maintenance. Eventually, you will need to replace them. Locate inverters for easy access for future service and replacement. Building codes may require less space than what you should actually design for. Always allow more space than code requires.
- Choose equipment for long life, modularity of components, and warranties. You should be able to rebuild equipment over time. Think about the manufacturer's information on replacing parts. Rebuilding parts may be an option. Understand how the manufacturer plans to support the components in their inverters.

Balance of System

Having a balance of system (BOS) involves making good decisions when selecting components for the PV system. BOS also pertains to how parts are installed and how they work together. With respect to BOS, here are some points to keep in mind:

- Opt for products that require fewest possible feet of conductor and the fewest possible connections. Conductor and connections create energy loss in the system. Limiting them reduces energy loss. Reducing the number of connections also reduces the risk of loose connections or inadequate tightening of a component.
- Install combiner boxes so that they are easy to access. This makes them easier to service and test. Design them to allow sufficient room in which to work. Having space to make solid connections makes installation and maintenance easy.
- Use smart combiners as a solution for string management and system monitoring. This is helpful in locations where service and testing are difficult. It will help in monitoring the PV system at the string level and maintaining expected performance levels.
- Mount and install hardware in such a manner that it will require little if any maintenance. For example, make sure brackets are heavy enough to ensure stability of the arrays. Double-check bolts to be sure everything is good and tight.
- Conduit can be oversized to assist in wire pulling. Oversized conduit can help with temperature reduction, especially when conduit is in the sun. Use plastic conduit except in locations with high temperatures. Protect plastic conduit if it is located in high-UV areas.

Wire Run and Sizing, Wire-Run Locations, and Protection

Conductors are the most important component of your PV system. You should design them with safety in mind. You should size them to meet or exceed the NEC.

PV design should reduce electrical losses. Losses are caused by temperature, connections, and conductor length. You should design and install PV systems with a resistance loss of 1 percent or less. Energy production increases when resistance is reduced in the PV system.

Initially, you should size wire and conduit according to the NEC requirements. Then adjust for temperature. Enlarge the wire by one full size for temperature, and then size to no more than about a 1 percent loss. This process will generally net a 1 percent loss or less. You will recover the additional wire cost in a year or two.

Performance will improve during the system's life. Select the right type of wire for the long-term life of the system. As electricity inflates in cost, oversizing the wire produces greater value than installing smaller wire that meets basic code design.

The following are some pointers in sizing and installing wiring:

- Reduce wire runs wherever possible.
- Keep conductor runs in the coolest location to reduce resistance losses.
- Address aesthetic issues in wire-run design. Communicate to the client any trade-offs for performance.
- Oversize the conduit by one size to make wire-pulling easier and to reduce conductor temperatures.
- All insulated conductors must be out of the sunlight and other environmental conditions.
- You must carefully design all conductors and conduit for excessive expansion and contraction to ensure wire protection. This includes all bends, joints, and interfacing connectors, which may separate or loosen during expansion and contraction. Stressing, stretching, or compressing the wire can lead to insulation failure and possibly a short circuit.
- All conductors and conduit must be designed to consider changes in the environment, which could compromise protection. This includes environmental and human-produced conditions that may cause damage.
- You should provide additional labeling for both visible and not-so-visible identification (for example, with conductors hidden behind walls or other construction materials).
- Always provide accurate as-built drawings that indicate exact details of the system. This is important for locations and design planning, and any issues related to phasing or expansion.

- Survivability of the conductors is a combination of design and installer responsibility.
- Survivability is for the life of the system, not just the local requirement of two to five years.
- Survivability and safety must include the use of wire, conduit clips, and wire ties that do not break down in local environmental conditions. Overcome time and elements in the design phases. Clips must last for at least 20 years.

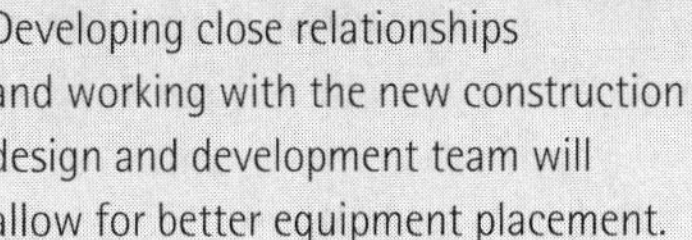

Developing close relationships and working with the new construction design and development team will allow for better equipment placement.

Equipment Burn-in Response

You must perform system checks after the burn-in period. Most PV systems have defects due to design and installation. Defects become clear in about 90 days. You should repair any underperforming panels or strings, inverter combiner boxes, switches, or compromised connections. This will improve the system's performance. Making repairs immediately will increase system output and reliability.

As of the end of 2010, most PV systems still do not include any monitoring systems (other than what is on the inverter). That means you know there is a problem with the system only if the system shuts down completely, the red light indicates the system is off, there is a major fault, or the customer sees a substantial jump in utility usage.

When performing the burn-in inspection, test for string balance, determine whether there are defective products, and check for the right torque on conductor and mechanical connectors. Performing these tasks may increase the PV system's performance by 2 to 5 percent. It also ensures that those connections are likely to maintain their effectiveness much longer than if they had not been serviced.

CHAPTER 2 SUMMARY

This chapter continued to explain the 36 criteria necessary to design a PV system that will fulfill a client's needs. It details the importance of correct panel and string sizing as well as how the NEC and local codes affect PV installation. Clear and concise communication between the PV design team and the client is essential to the success of the PV system. Communicating clearly with inspectors is important in obtaining certificates of operation.

It is important to design enough space in the PV system to perform maintenance and cleaning tasks. Keeping the PV system clean and well maintained will increase PV system output and increase components' lifespan. Installing oversized wiring and conduit can help increase PV system life and output.

KEY CONCEPTS AND TERMS

I-V curve

Maximum power point tracking (MPPT)

CHAPTER 2 ASSESSMENT

Assessing the Evaluation and Design Criteria

1. Inverters should be sized at 80 percent of the array's capacity.
   - ❑ **A.** True
   - ❑ **B.** False

2. Which two of the following are important to consider for string sizing? (*Select two.*)
   - ❑ **A.** System design
   - ❑ **B.** The structure to which the panels are attached
   - ❑ **C.** All environmental conditions
   - ❑ **D.** The material used to connect the panels to the structure

3. Conduit can be oversized for what two reasons? (*Select two.*)
   - ❑ **A.** To balance the system
   - ❑ **B.** To aid in wire pulling
   - ❑ **C.** To control the temperature
   - ❑ **D.** For aesthetic purposes

4. Which of the following can be found in the cavities of PV systems?
   - ❑ **A.** Inverters
   - ❑ **B.** Animals
   - ❑ **C.** Reptiles
   - ❑ **D.** All of the above

5. Cleaning a PV system can increase PV system performance by how much?
   - ❑ **A.** 15 to 20 percent
   - ❑ **B.** Approximately 2 percent
   - ❑ **C.** 10 to 15 percent
   - ❑ **D.** 5 percent or better
   - ❑ **E.** None of the above

6. Which of the following is the best direction for panel orientation?
   - ❑ **A.** A southwest direction
   - ❑ **B.** True south
   - ❑ **C.** True west
   - ❑ **D.** True north

7. The inverter needs to be efficient, even though no loads are using energy.
   - ❑ **A.** True
   - ❑ **B.** False

8. Transformerless inverters:
   - ❑ **A.** do not need to be grounded.
   - ❑ **B.** do not produce the additional heat that an inverter with a transformer does.
   - ❑ **C.** are said to be 98 percent efficient.
   - ❑ **D.** All of the above

9. Why should the PV system be kept clean? (*Select two.*)
   - ❑ **A.** To keep UV rays from damaging the modules
   - ❑ **B.** To reduce heat on system components
   - ❑ **C.** To increase system life
   - ❑ **D.** To reduce system performance

10. Tracking filter condition and soiling levels with accurate records is important.
    - ❑ **A.** True
    - ❑ **B.** False

Determining the Size of a Photovoltaic System

WHEN DESIGNING A PV system, you must make decisions about the type, size, technology, and placement of the of PV system. This chapter explains how to size a photovoltaic (PV) system. The size of the system depends on the client's energy demands, system goals, and objectives. Each client's needs, site, and energy consumption are unique.

The purpose of the PV system is also important. For example, a PV system supplying energy for a home will have different requirements from a PV system for a workshop or a manufacturing facility.

Topics & Concepts

This chapter covers the following topics and concepts:

- Determining the client's wants, needs, goals, and objectives
- Determining electrical demand
- Load profile
- Sizing standalone PV systems
- The power conditioning unit
- Sizing a grid-connected system with battery backup
- Sizing grid-tied PV systems
- Sizing the inverter

Goals

Upon completion of this chapter, the student will be able to:

- Assess electrical demand
- Discuss characteristics and functions of power conditioning units
- Construct a load profile
- Size a standalone system
- Size a grid-connected system

Determining the Client's Wants, Needs, Goals, and Objectives

To develop a successful client relationship, the PV team must understand the client's wants, needs, goals, and objectives. Communication with the client is important. This is much more complex than simply determining the client's electrical demand.

Factors such as budget, visual appeal, personal preferences, and the client's personality are important aspects. These factors are crucial in meeting the client's needs and expectations. The No. 1 skill in achieving success is to listen and understand exactly what the client wants!

Save time by communicating successfully with the client throughout the project. Understanding the client's needs and wants at the beginning of the project saves time during the design and installation phases. Educating the client at the beginning of the process reduces problems along the way. Work will proceed more quickly, and the client will be happier with the finished product—which is the goal. An unhappy client can destroy the reputation of a business. A happy client helps build it. After identifying a client's wants, needs, goals, and objectives, you must educate the client by conveying realistic options.

"A client/customer is the most important visitor on our premises. We are not doing him a favor by serving him. He is doing us a favor by giving us the opportunity to serve him."
—Author unknown

Determining Electrical Demand

Every client site has different energy, or load, requirements. Whether the site is a home, office building, manufacturing facility, workshop, or retail store, a PV system can reduce or eliminate the client's reliance on the current electrical grid. You must determine energy requirements at the outset of the project.

Grid-Tied System

In grid-tied systems, it is best to start with at least one year of utility bills (two years is better). You must then perform an accurate system analysis, which uses that data to determine an annual load profile and a monthly load profile.

The load profile will include the following client information:

- Name
- Address
- City, state, postal code
- Telephone numbers
- Parcel and lot numbers
- Distance from shop
- Number of stories of building
- Wire runs
- Roofing types, slopes, and related issues
- Usable areas for installing panels on the roof
- Maximum DC wattage on the roof
- Solar orientation factor(s)
- Solar performance factor(s)
- Proposed DC wattage
- Proposed AC wattage
- Utility rebate
- Federal tax credits, deductions, or grants
- State tax credits, deductions, or grants
- Energy inflation rate for the year
- Projected energy inflation rate for the future

Analysis Output

Analysis output is composed of all the factors needed by the client and the PV system designer to determine the impact the system will have on annual utility consumption and costs. The system should offset 50 percent or more of the total annual load. This will provide the client with some months with a zero energy bill for offset kilowatt-hours. If the system can't carry at least half of the energy load at the site, the client may not see any value to the system.

Off-Grid Standalone Systems

Understanding the client load profile is difficult in off-grid standalone systems. Create a load profile to determine the size of the system. The system must supply electricity for appliances, computers, power tools, and other types of equipment. The load profile identifies devices that have large energy loads.

You can shift large loads to another energy source. Alternative sources are propane, natural gas, a diesel generator, a small wind-powered system, or a small/micro hydro system. Shifting high-use devices such as refrigerators, water heaters, and freezers to other energy systems may help reduce the size and cost of the PV system. You can increase the size of the PV system later through careful system-phasing designs.

Load Profile

There are many considerations to take into account in creating the load profile:

- The electrical grid, natural gas, or both can supplement PV systems in urban locations.
- In rural areas, PV systems may have to be standalone.
- You can supplement PV systems with a generator and/or have battery backup.

A **load profile** lists all devices that use electricity to operate. You should list all devices with their wattage, how many hours they will run per day, and whether they use **alternating current (AC)** or **direct current (DC)**. You can find watts used in the manufacturers' literature or on the equipment plate. Calculate watts with the following equation:

Volts × amps = watts

You must know the wattage to calculate the average daily energy load the PV system will need to provide.

If the device uses AC, use an inverter to convert the energy from DC to AC.

The load profile will also identify devices that are candidates for load shifting. **Load shifting** is using other sources, such as liquid propane gas, liquid gas, or kerosene, to fuel devices that use electricity. These sources can fuel lights, hot water heaters, space heaters, refrigerators, and other appliances. In some cases, these devices are "dual fuel," which means they operate using electric and gas (or another fuel).

Interview the client in person to develop the energy profile. Interviewing the client will reduce and eliminate mistakes in designing and installing the PV system. The more information you can gather, the more accurate the work will be, and the less time it will take to create the finished product.

Here are a few tips to help reduce the load profile. Reducing the load profile reduces the energy demand on the PV system. A smaller load profile allows for a smaller PV system, reducing cost and space for the system. Expand the PV system as space and funds become available.

- Use Energy Star–rated appliances or appliances listed in the American Council for an Energy Efficient Economy.

- Shift the refrigerator load from using electricity by purchasing a refrigerator that runs on propane or kerosene. These fuels will inflate in price. Discuss this with the client.
- If an electric refrigerator is the only option, purchase one that uses DC and not AC. Alternatively; purchase a DC/gas model. This type of model is becoming more common for boats and RVs.
- Use a gas clothes dryer or a clothesline.
- Avoid using electric space heaters.
- Use compact fluorescent lights, which are much cheaper to operate than incandescent lights.
- Use a gas or propane stove instead of an electric stove.

Other factors to consider in determining the load profile, which are especially critical in standalone systems, are as follows:

- **Duty cycle**—The duty cycle is the period when a device is actually using power. For example, when a refrigerator motor is running to cool the inside, the refrigerator is in a duty cycle. Devices controlled by a thermostat, such as an electric blanket or a stove, run in duty cycles.
- **Phantom loads**—A device that always consumes energy even though it is not in a duty cycle has a phantom load, also called a vampire load. It is almost impossible today to purchase any appliance or device that does not have electronics that operate in the background, even when the unit is off. Such devices include but are not limited to digital clocks, TVs, stereos, washers, dryers, devices with remote controls, and battery chargers.
- **Surge loads**—Devices that use more energy when they start up than during ongoing operation create a surge load. Examples of devices that have surge loads are power saws, pumps, and any type of motor. Starting a power saw uses more energy than operating the power saw. The initial draw can be determined if the manufacturer's locked rotor load can be determined.

> **NOTE**
>
> When estimating the load for a device that creates a surge load, triple the watts used when the device is running. If the manufacturer lists the device's energy use as 500 watts, estimate the surge load at 1,500 watts. To ensure that the PV system generates enough energy to accommodate surge loads, multiply the surge load by seven.

It is important to understand the daily energy usage of all devices the client will be using. The formula to calculate the daily profile load of a particular device is as follows:

$$\text{Item and quantity } x \text{ volts } y \text{ amps} = \frac{\text{watts} \times \text{hours per day} \times 7 \text{ days per week}}{7 \text{ days per week}} = Z \text{ watt-hours per day}$$

- List the individual load (i.e., light bulbs, stove, refrigerator, computers, clothes washer, clothes dryer, televisions, radios, stereo, and power saws).

- List the quantity of each load (i.e., number of light bulbs: 46, refrigerator: 1, stove:1, freezer: 1, computer: 2, clothes washer: 1, hot water heater: 1, televisions: 3, radios: 2, stereo: 1, power saws: 2).
- List the watts as AC or DC. You will need to evaluate which items require AC or DC, which items can draw from batteries, and which items will require an inverter.
- List how many hours per day the device will be used. When calculating hours used, use the time of year the device will be used the most hours per day. For example, artificial lighting will be used more hours in the winter than in the summer, when there is more natural light available.
- List how many days per week the device will be used. Clothes washers might be used every day or just a few days per week; power saws may be used just a few days total per year.
- Divide the use days per week by 7 days.

This equals the watt-hours by AC/DC used by a particular device on a daily basis.

In the case of an existing grid-connected structure, you can use the past year's monthly electric bills to verify the average daily load used. Average the monthly watt usage over a minimum of one and preferably several years. Most electric bills use kilowatt-hours; to convert them to watts, use the following formula:

Number of kilowatts $\times$ 1,000 = watts

Sizing Standalone PV Systems

A standalone PV system is not connected to the grid. The system supplies all the required energy. Standalone systems store excess solar energy in a battery bank. This energy is available when insolation is low—for example, on cloudy days or at night.

You should size PV battery systems according to the three lowest months of insolation and the highest load demand. The months of lowest insolation are typically November, December, and January. Keep batteries from becoming depleted during long periods of low insolation with a gas- or diesel-powered generator.

Consider the following six factors when sizing a standalone PV system:

- Determine electric loads by using the load profile.
- Determine the battery size and type.
- Determine array size and type.
- Choose a charge controller(s).
- Determine the inverter(s) size and type.
- Determine wiring requirements for the system.

The Power Conditioning Unit (PCU)

The **power conditioning unit (PCU)**, also known as an inverter, is classified by the waveform it produces. The output waveform depends on the conversion and filtering methods. This smoothes the waveform, eliminating spikes and unwanted frequencies that occur when the switching occurs.

Three of the most common waveforms are as follows:

- **Square wave PCU**—A square wave PCU produces a switched AC output with little output voltage control, limited surge capability, and considerable harmonic distortion. The distorted waveform generates additional heat in the inverter and in the electrical appliance you are powering, which shortens the life of the appliance while requiring additional energy to produce the same amount of power.
- **Modified sine-wave PCU**—A modified sine-wave PCU can handle larger surges and has less harmonic distortion in its output. The distorted waveform generates additional heat in the inverter and in the electrical appliance you are powering. As with the square-wave PCU, this shortens the life of the appliance while requiring additional energy to produce the same amount of power.
- **Sine-wave PCU**—A sine-wave PCU produces an AC waveform, which is comparable or superior to most utilities. A sine-wave quality inverter is an electronic device that provides a stepped sine wave with so many thousands of steps that it looks and acts like a true sine wave. In today's world, with the cost of inverters versus the results, there is seldom a reason to use anything other than a sine-wave inverter.

Power Conditioning Unit/Inverter Characteristics

You must understand the following before selecting a PCU for a system:

- **Power conversion efficiency**—Power conversion efficiency is the ratio of output power to input power of the PCU. Efficiency of a standalone PCU varies significantly with the type and amount of load.
- **Load**—You should size the PCU for maximum energy use during prolonged periods of low insolation. During these periods, the PV system will be less effective and require an outside energy source such as a generator.
- **Air temperature**—It is important for the PCU to be located in an area where the air temperature is as moderate as possible. Extreme hot and cold temperatures reduce the PCU's efficiency and life. Review the inverter temperature range and warranty conditions; using an inverter in excessive temperatures voids many inverter warranties.

- **Airflow**—Adequate airflow will help cool the PCU and keep it from overheating. This will extend the life of the unit and increase performance by reducing internal operating temperatures.

You test inverter conversion efficiency under the UL 1741 standard. It provides data at 25 degrees Celsius, the same temperature at which PV modules are tested. As with panels, temperature and airflow affect performance; however, this is complicated by load when addressing inverter design issues. An overloaded inverter reduces efficiency.

Reducing airflow raises internal inverter temperatures, reduces efficiency, and shortens inverter life. When temperatures rise above 25 degrees Celsius, and especially above the inverter's maximum temperature, efficiency varies. In many inverters, operating above this temperature reduces performance, efficiency, and inverter life.

It is not wise to overload an inverter. When a designer or installer overloads an inverter, places it in over-temperature conditions, and/or provides reduced airflow conditions, the result is reduced efficiency and a dramatic shortening of inverter life.

Overloading the inverter may reduce the initial cost per watt and overall cost of the PV system; however, an overloaded inverter will fail quickly. The client will not be satisfied, and ultimately business will be lost. It is less expensive in the long run to install properly sized inverters in the initial installation.

There is a challenge in the inverter and string-sizing process. Understand that code, not performance, drives string sizing.

- **Rated power**—Rated power is an indication of how many watts of power the PCU can supply during standard operation at the certified specification. The PCU-rated power should be at least 125 percent of peak load demand. This is called the 80 percent guideline or rule. This allows some growth in load demand for the system.

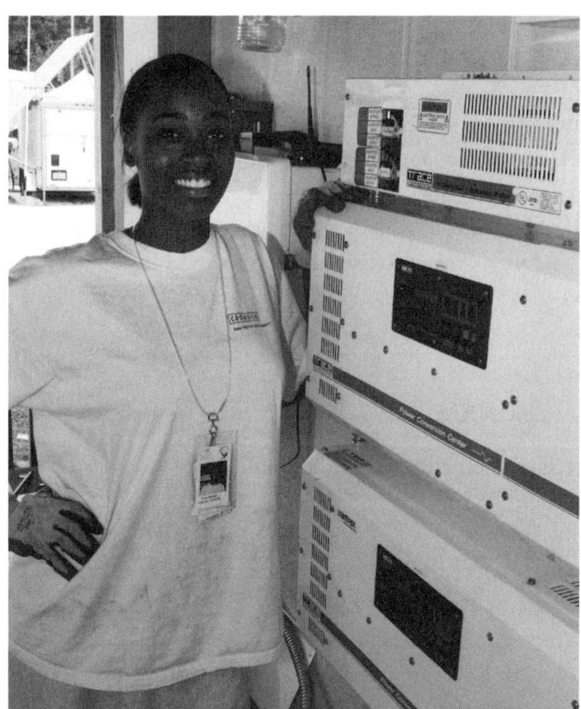

A Tuskegee University student works on the team's power-conditioning equipment to maximize efficiency.

Courtesy of DOE/NREL

- **Duty rating**—Duty rating is the amount of time the PCU can supply the maximum load. Exceeding this time may cause hardware failure, which is another reason the PCU should be oversized.
- **Input voltage**—Input voltage is determined by inverter specifications, inverter test standards, and code.
- **Surge capacity**—PCUs can exceed their rated power for limited periods, usually in seconds. Thus, you must determine surge capacity when designing off-grid systems for some loads.
- **Voltage regulation**—Voltage regulation indicates the variability in the output voltage. Better PCUs produce a nearly constant root-mean-square (RMS) output voltage over a range of loads.
- **Voltage protection**—A PCU can be damaged if DC input voltage levels are exceeded. Because battery voltage can far exceed nominal voltage if the battery is overcharged, it is advisable to select PCUs with voltage protection—that is, sensing circuits that disconnect the unit from the battery if either high or low levels are present at the input. Do this by using fusing, breakers, surge protection, or internal electronics in some inverters.
- **Frequency**—Knowledge of the load frequency is required in selecting a PCU. In the United States, most loads require 60Hz. In other parts of the world, 50Hz is commonly used. Variations in frequency can cause poor performance of clocks and electronic timers within the system.
- **Modularity**—It is sometimes advantageous to have multiple PCUs in some PV systems. You can connect these in parallel or use them to service different loads. The term for this is modularity. Modularity is also becoming a greater part of inverter design. Inverter components are modular, and can be replaced or rebuilt to extend the life of the inverter. This reduces its usable life cost. As a result, it is an important design decision to select the right inverter.
- **Power factor**—The power factor is the cosine of the angle between the current and voltage waveforms produced by the PCU. This value varies with the load. The power factor should be very close to 1.

Sizing a Grid-Connected System with Battery Backup

A PV system with battery backup tied to the grid has several advantages, one being that the system has two independent energy sources. With this type of PV system, the client's electrical service will probably never be interrupted.

In this configuration, the system has a subpanel supported by batteries. The client can choose which appliances, tools, and so on will use the battery backup portion of the PV system, and which will use the grid only. This requires selectivity and a lot of discussion and planning. It enables the client to begin with a smaller, less expensive PV system, and then add to it as funds become available.

In this scenario, you should install the more expensive nickel-iron (or some other form of more expensive, high-tech) batteries. The more expensive batteries will allow the owner to expand the battery system. If you do not install higher-quality batteries at the outset, the owner will not be able to expand the battery portion of the system without risking the whole battery array.

Sizing Grid-Tied PV Systems

A PV system connected to the grid is not required to provide 100 percent of the client's energy demand. This enables you to size the system according to the client's budget and financing restrictions, space limitations, and utility regulations.

To calculate the size of a grid-tied PV system, decide what percentage of annual electricity needs the client wants the PV system to provide. Divide the client's utility bill by the total kilowatt-hours used; then divide that figure by the percentage of energy the PV system will provide. Then, use this figure to calculate how many watts the PV system will have to produce.

For example, suppose a client uses 10,000 kWh per year and wants the PV system to provide one-third of that energy. 10,000 kWh/3 = 3,333 kWh; that is, the PV system will have to deliver 3,333 kWh per year. Now suppose the sun delivers 1,500 kWh per 1,000W of solar energy each year. So 3,333/1,500 kWh × 1,000W = 2,222 watts; that's how many watts the PV system needs to deliver. In this case, you should install a 3,000-watt inverter to allow for rounding, shading, voltage drop, and future expansion of the PV systeSizing the Inverter.

An inverter is an important component of the PV system. It converts DC to

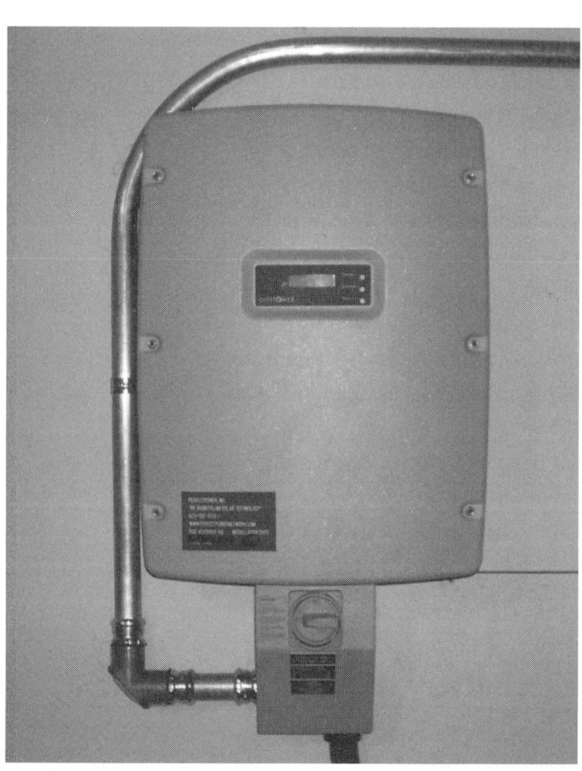

The inverter converts direct current to alternating current, which is used by most household electrical devices.
Courtesy of PerfectPower, Inc.

AC. The size of the inverter depends on how many watts it will have to convert from DC to AC, the estimated surge watts, and which type of waveform it will have to accommodate: square, sine, or modified sine. Consider only a good-quality sine-wave inverter; considering a lesser-quality sine wave inverter will only create operating problems in the PV system in the future.

To determine the size of the inverter, you need to know the following:

- Total AC watts
- DC system voltage
- Estimated starting/surge loads

The load profile lists the total AC and DC watts. When calculating surge loads, remember to multiply the watts used to operate the device by three to seven times. You can also determine watts used by using an ammeter.

A larger inverter will allow for later additions of energy requirements. For example, after you install the PV system, the client may add another computer, more lighting, an additional television, or another VCR. Oversizing the inverter gives the PV system flexibility. This enables the client to use multiple high-energy devices at the same time.

NOTE

When sizing any PV system, always allow for shading and voltage drop. Instruments are available to assist in determining the percentage of shade an array will receive under certain conditions. ASSET, SunEye, and White Solar Pathfinder are a few manufacturers of such tools. Consider voltage drop when sizing a PV system. Some solar installers allow for a 25 to 30 percent drop in voltage when sizing a PV system, and increase the system's size accordingly.

CHAPTER 3 SUMMARY

Evaluating and analyzing the client's energy load profile is one of the most important steps in determining the size of the PV system. The off-grid implications are that the load profile also assists the client in choosing appliances to run on electricity, natural gas, or propane. PV systems can be off-grid and supply 100 percent of the client's energy needs, or they can be connected to the grid and handle up to 100 percent of the client's energy requirements. You can design PV systems to allow for expansion as the client's energy needs increase or as the client's budget allows.

KEY CONCEPTS AND TERMS

Alternating current (AC)

Direct current (DC)

Duty cycle

Duty rating

Frequency

Input voltage

Load profile

Load shifting

Modularity

Phantom load

Power conditioning unit (PCU)/inverter

Power conversion efficiency

Power factor

Rated power

Surge capacity

Surge load

Voltage protection

Voltage regulation

CHAPTER 3 ASSESSMENT

How to Determine the Size of a Photovoltaic System

1. Load profiles list only large energy-consuming appliances such as refrigerators.
   - ❑ **A.** True
   - ❑ **B.** False

2. A standalone PV system requires which of the following?
   - ❑ **A.** A controller
   - ❑ **B.** An inverter
   - ❑ **C.** An energy-storage device like a battery bank
   - ❑ **D.** All of the above

3. Sizing the inverter depends on which of the following? (*Select two.*)
   - ❑ **A.** The axis of the array
   - ❑ **B.** Estimated surge watts
   - ❑ **C.** DC system voltage
   - ❑ **D.** The number of batteries the system requires

4. Power conversion efficiency depends on which of the following? (*Select three.*)
   - ❏ **A.** The tilt of the array
   - ❏ **B.** The load for battery backup systems
   - ❏ **C.** Airflow
   - ❏ **D.** Shading
   - ❏ **E.** Temperature

5. Grid-connected PV systems with batteries allow which of the following?
   - ❏ **A.** Clients to use the PV system during peak grid-demand times
   - ❏ **B.** Have energy available during blackouts
   - ❏ **C.** Stay dependent on the electrical grid longer
   - ❏ **D.** A and B only
   - ❏ **E.** None of the above

6. _________________ is the amount of time the PCU can supply the maximum load.
   - ❏ **A.** Duty rating
   - ❏ **B.** Modularity
   - ❏ **C.** Power factor
   - ❏ **D.** Surge rating

7. The power factor of the power-conditioning unit should always be greater than 1.
   - ❏ **A.** True
   - ❏ **B.** False

8. What are the three types of power conditioning units? (*Select three.*)
   - ❏ **A.** Sine wave
   - ❏ **B.** Modified sine wave
   - ❏ **C.** Square wave
   - ❏ **D.** Modified square wave
   - ❏ **E.** Conditioned wave

9. Direct current can be converted to alternating current by using a controller.
   - ❏ **A.** True
   - ❏ **B.** False

10. _________________ occurs when a PV system has the ability to replace or add components.
    - ❏ **A.** Modularity
    - ❏ **B.** Frequency
    - ❏ **C.** Power factor
    - ❏ **D.** Input voltage

Photovoltaic Array Configuration and Sizing

THIS CHAPTER INTRODUCES PHOTOVOLTAIC strings, array configurations, and sizing. The chapter begins with a brief overview of PV module fundamentals. Next, it reviews sizing the system and choosing the PV module. It then discusses array sizing along with a method for calculating PV array yields.

Topics & Concepts

This chapter covers the following topics and concepts:

- The systems approach to PV design
- PV module fundamentals
- System sizing and module choice
- PV array sizing
- System-modeling tools

Goals

After completing this unit, students are expected to be able to:

- Describe the basic principles, methodologies, and strategies for sizing photovoltaic systems according to the National Electrical Code (NEC) and for performance,
- Discuss system sizing and module choice,
- List module-selection criteria,
- Calculate the yield of a PV array.

The PV array is just one part of the overall system. Its size, design, and location will affect the rest of the system.
Courtesy of PerfectPower, Inc.

The Systems Approach to PV Design

The goal of a system concept is to lay a framework by which the following occurs:

- The PV team delivers a PV system that meets the client's needs and wants.
- The PV team employs a process that treats the installed product as an integrated system that will meet the project's goals and objectives.

Systems design is about two things:

- What the client wants to achieve
- How to deliver it through design, technology, and installation

System design involves the following goals:

- To deliver a system that generates a substantial amount of energy at a reasonable cost per kilowatt-hour
- For the system to be reliable
- For the system to have low operations and maintenance (O&M) costs
- For the system to be pleasing to the eye

Each project has many possible solutions. Finding the best solution takes time. Once the PV designer finds the solution, the designer must educate the client on how the components of the PV system work together. Having the components explained helps the client understand the value of the PV system.

PV Module/Panel Fundamentals

A PV module consists of photovoltaic cells. A **photovoltaic cell**, or PV cell, collects energy from the sun and converts it to DC electricity. The cells are usually about 1/100 of an inch thick. Each cell can produce half a volt or more of DC electricity.

There are many different types and qualities of PV cells and panels. You should learn about each type to make appropriate selections. Different types of cells have different functions and characteristics, based on composition and internal structure. The life span of a PV cell ranges from 10 to 50 years. The life span depends upon the type and quality of the cell and the local environment.

PV cells are assembled into a module, or **panel**. This protects the cells against degradation from environmental factors and weather. It makes them easy to work with. A group of modules is called a PV panel.

Modules are assembled and mounted on a frame to make a string. A string is a specified number of panels designed to provide specific ranges of voltage and current.

Strings are assembled to create an array. Strings in an array are wired together to produce a certain amount of voltage and current. Current from the modules is in the form of DC and voltage. The voltage then goes through an inverter. The inverter turns DC into AC.

> **NOTE**
>
> Professionals in the PV industry often use the terms "module" and "panel" interchangeably.

Voltage may vary due to temperature. Current may vary with the amount of sun available. Current is measured in watts or kilowatt-hours per meter squared (kWh/m²). A **watt** (W) is a unit of electrical power that occurs when a current of one ampere flows through a conductor with a potential one volt. One watt equals 1/746 of a horsepower.

Modules convert approximately 2 percent to 20 percent of the solar energy collected into electricity. The exact percentage depends on the type of module and the conditions in which the system is operating.

PV panels are tested and certified using standard test conditions (STC). STC allow for an instantaneous solar panel rating under controlled conditions. The use of STC is valuable in providing an international nameplate rating for PV modules. You must understand the limits of STC, however.

The STC for a module are as follows:

- Solar irradiance of 1,000W/m2, with the cell temperature maintained at 25 degrees Celsius (77 degrees Fahrenheit)
- Air mass of 1.5 at 0 wind speed for cooling effect

In a test, a panel is brought to temperature and exposed to a flash of artificial light to create the certification conditions. Panels rarely encounter these conditions in the real world, however. In addition, a PV system in Chicago, for instance, faces very different conditions from a PV system in Phoenix. In Chicago, solar irradiance for a fixed PV array with a south-facing tilt is about 60 percent of that in Phoenix. The irradiance numbers and slope are different.

You must consider this in the system design. Irradiance is less than $1,000W/m^2$. The temperatures for both the air and the panels have very different values. For comparison, the record high temperature in Chicago is 105 degrees Fahrenheit (on July 24, 1934), and the record-low temperature is -27 degrees Fahrenheit (in January 1985). In contrast, the record-high temperature in Phoenix is 122 degrees Fahrenheit (on June 26, 1990), and the record-low temperature is 16 degrees Fahrenheit (on January 7, 1913).

System Sizing and Module Choice

There are many types of modules from which to choose. The type of module chosen will have a significant impact on the size of the array necessary to produce the energy required by the PV system.

To select the appropriate module for a PV array, you must understand how the following four criteria affect the composition of the module:

- **Cells**—Cells affect the performance of modules. There are several types of cells, most of them made of **crystalline silicon**. Cell types include monocrystalline silicon cells, multicrystalline silicon cells, amorphous silicon cells, ribbon-drawn or pulled silicon cells, and thin-film technology cells. You'll learn about each of these in a moment.
- **Glazing**—Glazing is semitransparent and commonly used in non-view windows, curtain walls, facades, and skylights. Glazing can use either thin film or crystalline silicon cells. When integrated into a window, two panes of glass are used. With glazing, the PV module is placed inside the window's pane.
- **Electrical connections**—The location of positive and negative terminals on modules varies by manufacturer. Some manufacturers locate the terminals inside a junction box, also known as a J box. Other manufacturers use quick-connects, which have the positive and negative conductors already wired to the back of the module. These are much easier and faster to install.
- **Junction boxes, or J boxes**—Five to 10 years ago, all experienced PV installers made sure that panel conductors were kept inside conduit. The J-boxes were strung together with conduit from panel to panel and on to the combiner, if one was used. To relieve stress on the conductors, a cable clamp

was used when there was no conduit. With MC connectors and USE 2 cable, knockouts pretty much disappeared and if conduit was to be used, there was usually a threaded or PC connector used.

Monocrystalline Silicon

Monocrystalline silicon has an orderly atomic structure. This makes it predictable, with an efficiency rate of less than 15 to more than 21 percent. It is also easier to work with. The downside of monocrystalline silicon is that it is expensive to make. To create a predetermined and orderly cell structure, the manufacturing process is slow and precise. The development of monocrystalline silicon begins with the production of crystals of silicon, made from molten silicon. This silicon is very pure. The molten silicon forms a single crystal ingot, which is cut into thin slices. These slices usually range from 0.2 to 0.3 mm in thickness. Each wafer is a monocrystalline solar cell. The edges of the round wafers are cut off, creating a hexagonal shape, allowing more wafers to be placed on a module.

Multicrystalline Silicon

Multicrystalline silicon, also called polycrystalline silicon, is a semiconductor with a less orderly cell structure. It is less expensive to make than monocrystalline silicon. Carrier flows are blocked in multicrystalline silicon. This reduces cell performance and allows higher energy levels in the forbidden gap. The extra energy levels create quality recombination sites. This creates the paths needed for current to flow across the p-n junction. Multicrystalline silicon cells have an efficiency rate of 13 to 16 percent.

Amorphous Silicon

Amorphous silicon is the least expensive PV semiconductor to manufacture. Amorphous silicon has a very loose structure, which results in dangling bonds. Dangling bonds create extra energy in the forbidden gap. Atomic hydrogen can be used to saturate the dangling bonds. This improves the quality of the photovoltaic material. Amorphous silicon tends to have relatively low efficiencies, but is also the least expensive to manufacture.

Ribbon-Drawn or Pulled Silicon

Ribbon pulling is a solar-cell–manufacturing technique that significantly reduces waste. Instead of sawing the silicon into cells, ribbons of silicon are pulled from the silicon melt. The ribbons are already the necessary thickness. Eliminating the dust produced in the cutting process means less waste. Eliminating the need for sawing also results in the use of less energy. Ribbon-pulled technologies used in commercial solar cell production include the string ribbon process and edge-defined film-fed growth (EFG) methods. Both of these methods result in pulled wafers consisting of a strip, or ribbon, of silicon. APex technology results in polycrystalline thin-film solar cells on substrate.

Thin-Film Technology

Thin-film solar cells are easier and less costly to manufacture. Unlike crystalline cells, they are not restricted to standard solar-cell–wafer sizes. The substrate used for thin-film cells is usually glass. The idea behind thin-film solar cells is that the substrate can be any shape and size and can be coated with a semiconductor. Only same-size cells can be strung together, so the most common shape is rectangular. Thin-film cells are monolithically connected during the layering and coating processes. This differs from the soldering technique used for crystalline cells. Thin-film cells have a conductive oxide top coating and opaque metal coating on the back to make the electrical contacts. Instead of referring to thin-film solar cells in terms of cells and modules, thin-film cells are called raw modules. This is because the semiconductor material is in the substrate and laminated. Thin-film cells are not as efficient as other solar-cell types. However, thin-film cells are less susceptible to shading and low light, and have a better temperature coefficient.

Module Selection

You should base module selection on specifications provided by the manufacturer. Compare specifications, such as the following:

- Performance
- Physical size
- Cost between different modules

Then decide which module(s) to use.

Consider whether the module will require an inverter. Some module junction boxes have inverters installed inside. Other inverters are installed outside the module. Modules with quick-connects do not have accessible junction boxes.

Designers tend to select modules on efficiencies rated at STC. STC ratings are often misleading. If you use only STC ratings to select modules, the result will be subprime performance. Unanticipated losses from mismatching components can also result.

Avoid problems from using only data from STC in choosing modules. Understand the local environment as it relates to module characteristics. This is important in the following areas:

- Temperature coefficients
- Cell architecture

- Application
- Airflow

Have data from real systems to compare when modeling and making module selections. Comparing data confirms the technical choices.

Number of Modules

You must determine how many modules should be connected in series and parallel to produce the required voltage and current. This is necessary to figure the strings. String sizing requires correct data from local conditions.

Adjust data from the model and real conditions. Consider system and irradiance availability requirements. Series-connected modules produce the design voltage in the right ranges for the inverter's maximum power point tracking (MPPT) algorithm.

Understand how specific MPPT performance boxes can affect string sizing. This limits choices and raises performance. Any DC energy that falls outside the MPPT parameters can result in lost energy on the AC side of the inverter.

Photovoltaic Array Sizing

A photovoltaic (PV) array is a series of strings of solar modules connected by wiring to collect energy from the sun. Each module has a positive and negative output. Modules produce voltage for energy-using devices such as appliances, tools, lighting, water pumps, and signage. These devices are used in homes, offices, shops, and utility companies, and by manufacturing firms.

The following determines the power output of the PV array:

- Size
- Exposure to sunlight
- All environmental factors
- Location

The following affect the PV array's ability to collect the sun's energy:

- Hours per day of sunlight exposure
- Pollution
- Dust
- Temperature
- Cloud cover

You should locate PV arrays where they can collect the most unobstructed light from the sun. The energy collected by the PV array converts into electricity for use by the client. Trees, taller buildings, chimneys, antennas, and other objects can shade the array.

You should avoid shading on the array whenever possible. Shade reduces the amount of energy the array collects and the voltage it produces. Consider hours of sunlight and shading when sizing the array. An array's sunlight exposure can range from less than one hour of sun to 14 hours per day. You can determine average daily sunlight exposure by using a world sunlight exposure map. Both NREL and Sandia Labs provide excellent resources in this area.

Sizing the PV Array

The size of the PV array depends mainly upon the following:

- Daily or critical loads for standalone and battery backup grid-tied systems
- Yearly solar irradiation details
- Autonomy
- Orientation of the array
- Tilt of the modules
- Cell temperature
- Shading
- Energy losses that occur in the system

Use the preceding to calculate the following:

- The number of modules
- The number of inverters
- The number of batteries
- The final output of the PV array
- The economics of the system

You should base array sizing in grid-tied systems on how much clients want to reduce their utility bills. Budget is also a consideration.

The daily load and the string sizing also factor into array sizing. Compare different panels and inverters. As you make comparisons, relationships between loads and equipment will become obvious. The best choices will become evident. The voltage-input ranges of charge controllers and inverters will determine how many panels should be in each

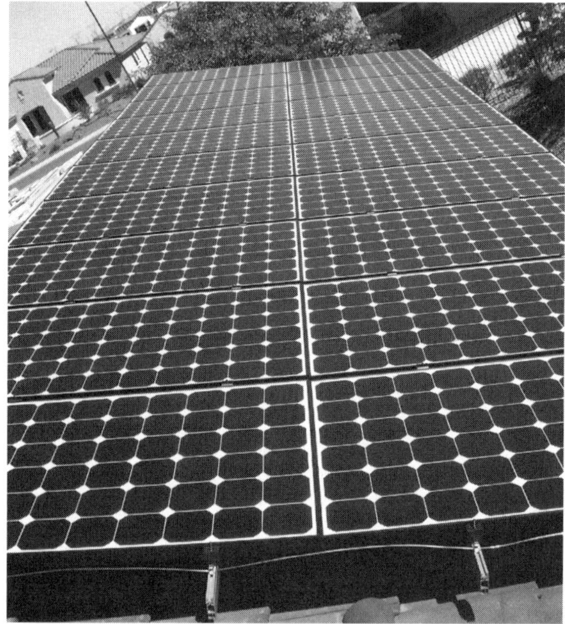

The required array size can be measured using a variety of methods including calculations and software programs. Most designers today prefer the use of software programs.
Courtesy of PerfectPower, Inc.

Standard outdoor measurement system: current versus voltage (I-V). The standard outdoor measurement system is our principal system for measuring module performance under prevailing outdoor conditions.
Courtesy of DOE/NREL, Credit: Jim Yost

string. This will help in determining the final array size. The standard outdoor measurement system is our principal system for measuring module performance under prevailing outdoor conditions.

Model for Calculating the Yield of a PV Array

Use the following to calculate the daily watt-hours for a solar-panel array. A **watt-hour (Wh)** is an energy unit of 1 watt of power operating for 1 hour usually, used in units of 1,000 (called a kilowatt-hour, or kWh).

$$P_{PV} = W/Z_2 \times Z_3 \times Z_4 \times V$$

- W is the daily load in watts.
- $Z2$ is the horizontal radiation. Use the month with the lowest insolation.
- Z3 is the PV array orientation minus the 45-degree tilt.
- Z4 is the cell temperature.
- V is the conversion, cable, and other energy losses.

Orientation

You should mount the PV array at a fixed angle from the horizontal, or on a sun-tracking mechanism. This raises energy production. If there are no obstacles, a tilt

MODULE VOLTAGE CORRECTION

One of the most critical design considerations for PV modules is the change in voltage due to temperature. All PV modules have an inverse relationship to temperature; as the module temperature increases, the voltage decreases and vice versa. This basic fact plays a part in all PV design and installations, and as a designer, you need to evaluate the change in voltage for both cold and hot temperatures.

The National Electrical Code (NEC) addresses the change in module voltage, but only for the "lowest expected ambient temperature" for the installation site. This is because low temperatures will increase the module voltage, and if this isn't checked, there is a risk of damaging components from excessive voltage. For many designers, using the lowest recorded temperature is the answer in this scenario. You can also consider using the data provided by the American Society of Heating, Refrigerating and Air-Conditioning Engineers (ASHRAE) for the mean extreme minimum dry-bulb temperature. (Dry-bulb temperature is what is usually thought of as air temperature; the "dry bulb" is a reference to the bulb of a thermometer freely exposed to the air but shielded from radiation and moisture.) Either temperature value can be justified and will result in proper system design.

The NEC doesn't address the effects of high temperatures on module voltages because that is considered a performance issue, not a safety issue. As a designer, though, you need to figure out how low the module voltage will go based on the expected summertime temperatures at the array location. This will allow you to know how many PV modules you need to place in series and keep whatever piece of electronics the modules are connected to running. To calculate this value, you can use the ASHRAE 2 percent summertime design temperature averaged for June-August. This represents a temperature that is exceeded only 2 percent of the time, resulting in a reasonable design criterion for your installations.

These temperature considerations will require that you run two separate calculations, one for low temperatures that will yield an increase in the module's open-circuit voltage (Voc) and the second that will yield a decrease in the module's maximum power, or operating, voltage (Vmp). It is important to apply the corrections to the proper voltages, since the array will start at Voc in the morning and will be operating somewhere off the Vmp in the afternoon.

To properly apply the calculations, you need to gather a few pieces of information from the module manufacturer and site-specific temperature data. For example, if an array of Evergreen ES-A-200 modules are located in Corvallis, Ore.:

Voc @ STC = 22.6V

Vmp @ STC = 18.1V

Temperature correction factor for Voc = −0.31%/C

Temperature correction factor for Vmp = −0.40%/C

Record cold temperature = −22°C

AHSRAE cold temperature = −8°C

AHSRAE hot temperature = 33°C

MODULE VOLTAGE CORRECTION (*Cont.*)

The next step is to estimate the temperature of the cells. For the low temperature, this will be the same as the cold ambient temperature selected. This is because first thing in the morning, the sun won't be able to heat up the array but there will be sufficient light to produce full voltage. The cell temperature in the middle of the day will be based on the local ambient plus some number of degrees based on the mounting method used. You can expect the modules to operate 30 degrees Celsius above ambient during the summer, and in some cases maybe even more.

Given this, you can estimate the cell temperatures based on the site data collected. If you want to adjust the module voltage based off the record cold temperature and ASHRAE high, the cell temperatures will

Tcell cold = –22°C

Tcell hot = 33°C + 30°C = 63°C

Now you can apply the numbers into the voltage correction equation:

$$V_{adj} = V_{ref} \times \{1 + [(T_{cell} - T_{ref}) \times \text{Temp Correction}]\}$$

Where:

Vadj = Adjusted voltage

Vref = Reference voltage at STC conditions

Tcell = Temperature of the PV cells in °C

Tref = STC temperature (25°C)

Temp Correction = Module voltage correction in % per °C

For the cold temperatures:

$$V_{adj} = 22.6V \times \{1 + [(-22°C - 25°C) \times -0.31\%/°C]\}$$
$$= 22.6V \times \{1 + [(-47°C) \times -0.31\%/°C]\}$$
$$= 22.6V \times \{1 + [14.57\%]\}$$
$$= 22.6V \times 1.1457$$
$$= 25.9V$$

For the hot temperatures:

$$V_{adj} = 18.1V \times \{1 + [(63°C - 25°C) \times -0.40\%/°C]\}$$
$$= 18.1V \times \{1 + [(38°C) \times -0.40\%/°C]\}$$
$$= 18.1V \times \{1 + [-15.2\%]\}$$
$$= 18.1V \times 0.848$$
$$= 15.3V$$

Now that you have the adjusted voltages, you can verify the number of modules in series needed to keep your equipment operating in all conditions. You will divide the maximum allowable voltage by the adjusted Voc and round down to the next whole number to determine the maximum number of modules in series. And for the lowest operating voltage, divide the equipment's lowest acceptable voltage value by the adjusted Vmp and round up to determine the fewest you can have in series.

For example, for an inverter that can accept up to 600V and no less than 250V, and using the Evergreen modules:

600V/25.9V = 23.2 or 23 modules maximum

250V/15.3V = 16.3 or 17 modules minimum

—Ryan Mayfield, president of Renewable Energy Associates, Corvallis, Ore.

PV modules mounted by PerfectPower in Phoenix, Ariz., showing module orientation.
Courtesy of PerfectPower, Inc.

angle at or near the location's latitude will give the most energy on an annual basis. Tilt angles of latitude plus or minus 15 degrees will increase energy production in winter and summer, respectively **FIGURE 4-1**.

Central Inverter Concept

PV cells, modules, strings, and arrays produce electricity in the form of direct current (DC). The array captures energy in times of high insolation. A **charge controller** stores the energy in a battery bank or on the grid. Storing energy makes it available during periods of no insolation.

When voltages match, you can run DC-powered devices directly from the PV system. Most electrical devices are manufactured to use AC, however. For this reason, PV systems need an inverter to convert DC to AC. If the PV system is grid-tied, the inverter will automatically mirror the AC waveform produced by the grid. Most PV systems use a single inverter, although some systems do use more than one (although the number of inverters used is limited).

Module Inverter Concept

Instead of using a single or a limited number of inverters, PV designers sometimes use micro inverters. When a system uses micro inverters, it means that each module has its own micro inverter. There are advantages and disadvantages to using micro inverters in a PV system.

One advantage of using micro inverters is that you mount them on the back of each module, which reduces energy loss. Matching modules and their inverters together is critical to ensure maximum efficiency of the module and the PV system. The micro inverter is specific to the module, which may give the PV designer greater flexibility.

Because there is one inverter per module, strings size is limited to one. This can help address shading issues on complex roof structures or where it is difficult to assemble coherent strings. In addition, you can monitor micro inverters on each individual module. Because you can individually monitor each module, it is easy for the client or the monitoring company to identify a troublesome panel.

There are some disadvantages to micro inverters. Because each panel has its own micro inverter, there are more inverters that can fail. More micro inverters also mean more parts and more connections. This makes the PV system complicated. Complicated systems are more likely to fail than simple systems.

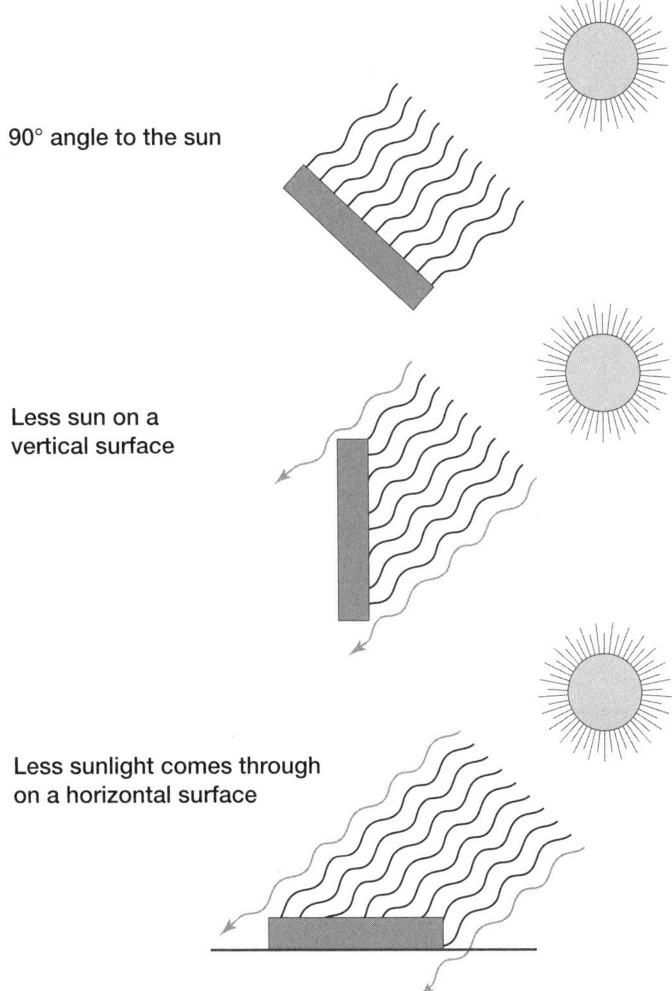

FIGURE 4–1 Module tilt.

Micro inverters solve shading issues, but reduce the panel's efficiency by 25 to 50 percent. It can be difficult to fit all the micro inverters in the system. You should locate micro inverters where they will be easy to replace if necessary. Another disadvantage of the micro inverter is that it requires more monitoring than other parts of a PV system. If the PV system owner chooses, a monitoring company can be hired to take care of this, although this would create additional expense.

Software Tools

String-sizing and system-modeling computer software is an invaluable tool for maximizing the efficiency of the PV system. Software is available to assist the designer in planning, sizing, and then simulating the final PV system. Using a computer simulation for a potential PV system design enables you to detect inconsistencies caused by inverters and correct mismatched modules before the design is complete and installation begins. You can compare different module and inverter configurations and choose the best possible outcome. You can define the operating behavior of the PV system, including how it reacts to the following:

- Shade
- Weather
- Orientation
- Tilt of the modules

To get the best results from the software, determine what the appropriate data to input should be, understand the limitations of the software, and understand what the data says. String sizing and system modeling can be very valuable. However, you must be able to confirm that the model is accurate as you input various design assumptions into the program.

Be aware that most software does not adequately address the temperature and MPPT algorithms and other issues. As a result, the program may give skewed results. It is important to double-check the data that you enter and to verify the results with a colleague.

Some inverter manufacturers provide free software for designers. If you look around on the Internet, you should be able to find a number of simulation software programs to help design, size, and implement an off-grid or grid-tied PV system.

Modeling software is a great tool. It is particularly helpful when you discuss the PV system with the customer. Simulations provide computer graphics that help the customer understand the value of the PV system. This understanding helps the client make good decisions about the system. Keep graphics simple when using them to explain the PV system to the client.

> **NOTE**
>
> Generally speaking, the more accurate the software, the more difficult it is to master. This effort usually is rewarded with greater accuracy, however.

System-Modeling Tools

The charts, tables, and worksheets will assist the PV designer in determining the size of the system:

- Load calculation worksheet
- Battery worksheet

- Array sizing worksheet
- Inverter sizing worksheet
- Controller worksheet

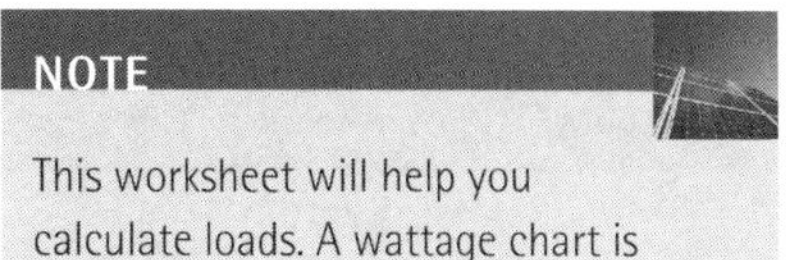

TABLE 4-1   LOAD CALCULATION WORKSHEET

Load Item	Number ×	Volts ×	Amps =	Watts ×	Use ×	Use ×	÷ 7 =	Watt-hours
				AC DC	Hours/ day	Days/ week	Days	AC DC

TABLE 4-2   BATTERY SIZING CALCULATIONS WORKSHEET

Daily AC average load Wh/day	÷	Inverter efficiency	+	Daily DC average load Wh/Day	÷	System DC voltage	=	Amp hours/ day average
Amp Hours/Day Average	×	Days of autonomy	÷	Discharge limit	÷	Battery capacity AH	=	Parallel batteries
DC Voltage (System)	÷	Voltage battery	=	Series batteries	×	Parallel batteries	=	Total batteries

TABLE 4-3   ARRAY SIZING CALCULATION WORKSHEET

Amp hours/ day average	÷	Battery efficiency	÷	Hours/day peak sun		Peak amps array		
Peak amps array	÷	Peak amps module	=	Parallel modules		Short- circuit current module		
System voltage DC	=	Module voltage nominal	=	Module in series	×	Modules parallel	=	Modules/ total

TABLE 4-4   INVERTER SIZING CALCULATION WORKSHEET

Total connected watts AC	÷	System voltage DC	=	Surge watts estimated	Desired features

TABLE 4-5   CONTROLLER SIZING CALCULATION WORKSHEET

Short-circuit current module	×	Parallel modules	×	1.25	=	Short-circuit amps array	Array amps controller	Desired features
	×		×	1.25	=			
Total DC connected watts	÷	System voltage DC	=	Maximum load amps DC			Load amps controller	

CHAPTER 4 SUMMARY

This chapter described the basic principles, methodologies, and strategies for sizing a PV system. Cells made of crystalline silicon are connected together into a module, or panel, and capture the sun's energy. Modules are installed at specific angles and are tilted to collect the maximum sunlight available.

Energy from the sun is collected in the form of DC power. Inverters convert DC into AC. You can install inverters on the back of the module or centrally within reach of the PV system array.

The size of the PV system depends on the daily load, modules used, design techniques, loss of energy due to inverters, cell temperature, batteries, and more.

Computer software is available to assist with modeling the PV system. This software can help size the PV system and simulate how the final PV system will function. This tool not only helps the designer to design the most efficient PV system, it helps the customer understand what he or she is purchasing.

KEY CONCEPTS AND TERMS

Charge controller

Crystalline silicon

Panel

Photovoltaic cell

Watt (W)

Watt-hour (Wh)

CHAPTER 4 ASSESSMENT

Overview of Advanced Photovoltaic System Design and Design Criteria

1. An array is a series of solar modules and strings.
   - ❑ **A.** True
   - ❑ **B.** False

2. The power of the array depends upon the array's:
   - ❑ **A.** size.
   - ❑ **B.** exposure to the sun.
   - ❑ **C.** location.
   - ❑ **D.** All of the above

3. Which of the following two substances are used to make PV cells? (*Select two.*)
   - ❑ **A.** Crystalline silicon
   - ❑ **B.** Silica fume
   - ❑ **C.** Silica gel
   - ❑ **D.** Amorphous silicon

4. Which of the following three are parts of an array? (*Select three.*)
- ❑ **A.** Shading
- ❑ **B.** Module
- ❑ **C.** Cells
- ❑ **D.** Insolation
- ❑ **E.** Inverter

5. An inverter does which of the following?
- ❑ **A.** Converts moonlight to electricity
- ❑ **B.** Flips cells in the module when they are full
- ❑ **C.** Regulates the amount of energy collected by the PV system
- ❑ **D.** Converts DC power to AC power
- ❑ **E.** None of the above

6. Devices powered by DC can be run directly from the PV system without an inverter.
- ❑ **A.** True
- ❑ **B.** False

7. A photovoltaic array can be mounted on a fixed angle from the horizontal in order to increase energy production.
- ❑ **A.** True
- ❑ **B.** False

8. A(n) _______________ is a device that controls the rate and/or state at which batteries receive charge from the modules.
- ❑ **A.** inverter
- ❑ **B.** array
- ❑ **C.** wiring
- ❑ **D.** charge controller
- ❑ **E.** computer software module

9. Which of the following are standard test conditions for a module? (*Select three.*)
- ❑ **A.** Solar irradiance of $1,000 W/m^2$
- ❑ **B.** Cell temperature maintained at 25 degrees Celsius (77 degrees Fahrenheit)
- ❑ **C.** An air mass of 1.5 at 0 wind speed for cooling effect
- ❑ **D.** Solar irradiance of $1,000 W/m^3$
- ❑ **E.** An air mass of 1 at 0 wind speed for cooling effect

10. Each cell can produce 0.5 volt or more of DC electricity.
- ❑ **A.** True
- ❑ **B.** False

Mounting Systems

THIS CHAPTER EXAMINES THE pros and cons of several mounting options. This chapter also includes ways in which various mounting options affect the system.

Topics & Concepts

This chapter covers the following topics and concepts:

- Mounting system functions
- Mounting system options
- Pole mounts
- Rack mounts
- Tracking systems
- Building-integrated photovoltaic (BIPV) systems

Goals

After completing this unit, students are expected to:

- Describe the basic principles, methodologies, and strategies for sizing photovoltaic systems to code
- Understand the function and role of mounting systems
- Discuss system-mounting options
- Describe the role of mounting within the PV system

The mounting system supports the PV panels and also directs them at the optimal angle.

Courtesy of PerfectPower, Inc.

Mounting System Functions

In terms of mounting PV modules, there are several factors for the PV designer to consider:

- **Location**—You should locate modules where they may capture the majority of the sun's energy. Modules must face south in North America.
- **Shade**—You should locate modules where they are not subject to shade.
- **Power conditioning unit (PCU)**—You should locate modules near the PCU to minimize the loss of electricity.
- **Maintenance**—You should locate modules in such a manner that you can easily perform maintenance on the system without interfering with the building or structure maintenance.
- **Aesthetics**—You should locate modules so they blend with the structure and are pleasing to the eye.
- **Theft/vandalism**—In some locations, you must protect the PV system from theft or vandalism. Not just people but animals can pose a physical threat to PV systems. For example, if the PV system is located in a field with herd animals such as goats, cattle, or horses, you should install a barrier such as fencing around the system.

Mounting System Options

There are several options to consider when choosing where and how to mount the PV modules **FIGURE 5-1**. For example, you could integrate the array on the roof or as a shading device over a porch when locating modules for a residence. For a commercial location, you might locate arrays over the parking area. You must determine if the desired support structure such as the roof is strong enough to support the weight of the system, especially the modules. In some cases, you may need to consult with the roofer or an engineer to determine structure strength. In all cases, you should carefully examine all potential mounting locations.

If mounting to a structure is not an option, you should consider mounting the array to a pole. Wherever you locate the modules, it should allow for easy access for maintenance purposes.

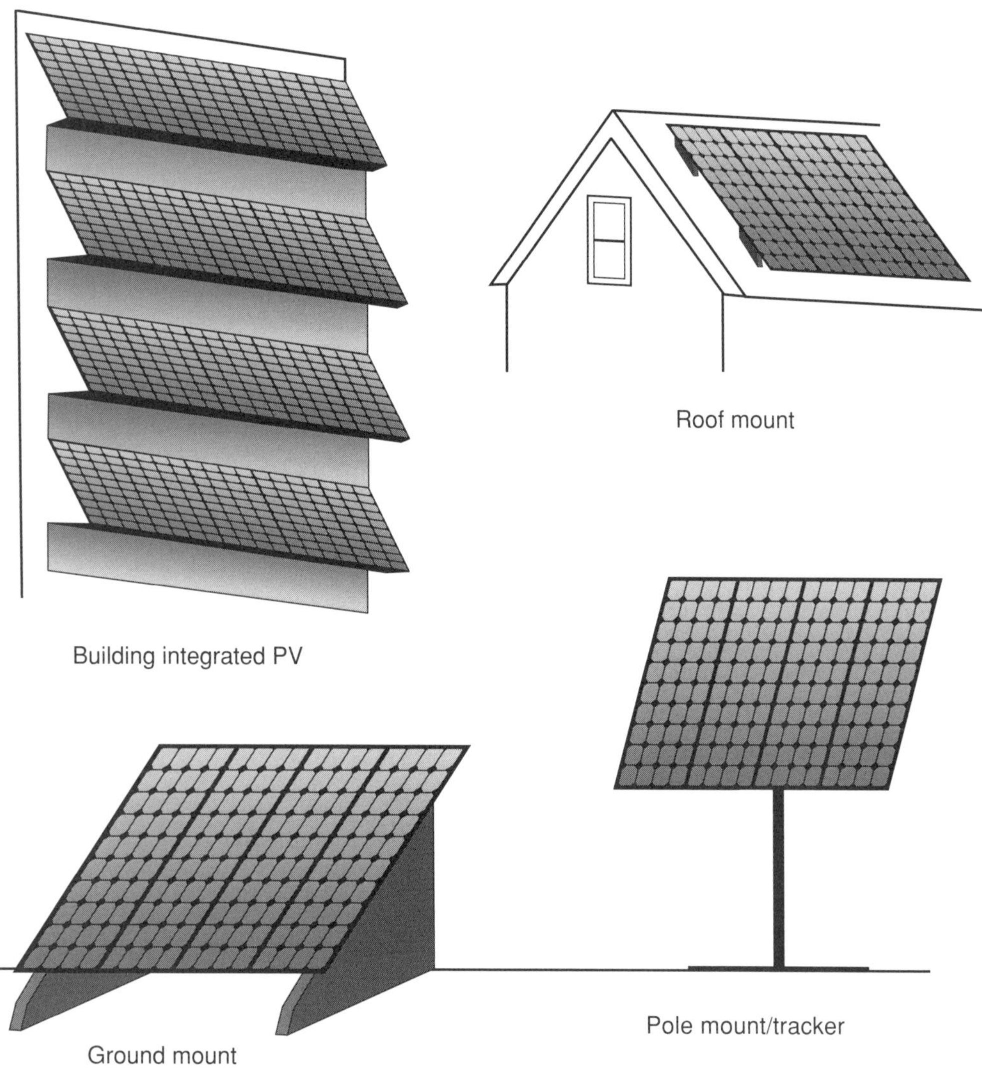

FIGURE 5–1 Mounting options.

Consider space when deciding where to mount the arrays. Each square foot of a PV system produces between 5 and 180 DC watts of electricity. (This output is contingent upon the quality of the components, shading issues, and all the other environmental issues that go into design.) An additional 20 percent of space is often required for accessing the system for maintenance and repairs.

Pole mounts are often stabilized by a heavy concrete base. The larger the array, the larger the base that is required.
Courtesy of PerfectPower, Inc.

Pole Mounts

Mounting the PV arrays to a structure is not always practical. One alternative is to mount the PV system on a pole. You should use pole mounting in rural areas, where structures are not available. For example, a PV system used for a water pump in the middle of a field may require a pole mount. The pole size determines the amount of cement needed. The size of the pole depends on the size of the system installed.

You should consider wind speed when using a pole to mount a PV system. Additional anchoring is required if the site experiences high winds. Additionally, you should determine terrain and soil type when installing a pole-mounted PV system. Contact a local geologist to assist with determining the soil type and possibly suggesting how deep to anchor the pole into the ground. Rocky soil presents difficulty in cementing poles.

Pros and Cons

There are many positive aspects of using pole mounts for PV arrays:

- You can situate arrays away from shade-producing objects.
- You can position arrays to face south and tilt them at just the right angle.
- You can install arrays more safely on a pole than on a roof.
- You avoid the need to bore holes in the roof for mounting devices and wiring.
- Pole-mounted arrays are cooler than roof-mounted arrays.
- Because pole-mounted arrays are cooler, they produce more output, thus potentially lowering the cost per kilowatt-hour.
- Locating arrays on poles makes access to the PV system easier, thereby making maintenance easier.
- It is often easier to clean dust and snow from a pole-mounted array due to better access.

There are also a few disadvantages to pole-mounting PV arrays:

- If arrays are too far away from the PCU, the system will experience a higher transmission loss. That means more modules will be required to compensate for the loss of output. You can offset this by increasing the DC wire size, but this may result in unwanted tradeoffs.

- Wind creates a disadvantage for pole-mounted arrays. You must design for the highest historical sustained wind speed experienced in the area to prevent wind damage to the array.
- The threat of vandalism means that urban areas are usually not good locations for pole-mounted systems. In addition, in urban areas, it might be difficult to locate the array away from shade-producing structures such as buildings, antennas, trees, or other objects.

TECH TIPS

If the array is located on a pole in a field, install a fence around the system to protect it from animals.

Rack Mounts

Using a rack is a popular option for mounting PV systems. Racks are either fixed or adjustable. Fixed racks are installed at a permanent, fixed angle. You should install fixed racks at an angle matching the degrees of latitude of the location whenever possible to maximize annual output, e.g., 33.5 degrees for Phoenix to match 33.5 degrees north. Some adjustable racks have telescoping legs; other racks feature predrilled holes to adjust the angle of the array for uneven surfaces or a variety of predetermined slopes. In the past, most racks were made of steel. Now, racks are generally made of aluminum to lighten the load on the structure and reduce corrosion.

Racks have three main components:

- **Anchor/feet, mounting brackets, or standoffs**—These attach the rack to the surface of a building, the roof, or a foundation on the ground.

This rack mount is attached to a sloped roof facing south.
Courtesy of PerfectPower, Inc.

SEASONAL ADJUSTMENTS WITH RACK MOUNTS

You cannot use a rack's legs to change the tilt of the rack for seasonal adjustments. In fact, designing a system that requires seasonal adjustments, typically performed twice per year, is greatly discouraged for the following reasons:

- Customers usually quit making the adjustments after a year or so.
- New owners seldom know about the adjustment process.
- Panels weigh a lot, and making seasonal adjustments involves a high risk of damage.
- Mechanical connections are seldom if ever torqued. This damages the rack over time, because incorrect tightening puts stress on the hardware.
- Most PV mounting equipment is not designed for seasonal adjustment, even when it is assumed that it is.

- **Bars/rails**—These hold the modules.
- **Legs**—These attach the PV system to the bars.

Racks are easy to install and can be attached to buildings, roofs, or the ground. When attached to a building, the PV system not only provides electricity, it also provides shade. For example, PV arrays can be installed as roofing over a garage or as an awning to provide shade over a deck or patio.

Pros and Cons

Mounting PV arrays on racks has its advantages:

- You can mount racks on almost anything. This includes awnings, roofs, or sides of buildings.
- Racks installed on roofs allow air to circulate more freely around the modules. Moving air helps cool the modules and increases output of the module.
- Mounting a rack on a roof enables you to install the array at a height greater than that of surrounding buildings and trees. This helps prevent shading on the arrays.
- Racks hold a variety of different-sized modules.

Racks also come with disadvantages:

- To install a rack on a roof, you must penetrate the roof at several points. You must adequately seal each point of penetration so the roof does not leak.

- High winds pose a problem for racks installed on roofs. Extra anchoring may be required.
- Typically, a PV system will outlive the roof on which it is installed. If the roof needs replacing, the racks must be removed.
- It can be difficult to install a rack if the roof is steep. This increases the cost of installation.

Tracking Systems

Another way to mount PV system arrays is to use a tracking system. There are two types of tracking systems:

- **Active tracking systems**—These are systems with motors, powered by their own small PV panels, to move the arrays.
- **Passive tracking systems**—These systems don't have motors, gears, or controls. They include a refrigerant, which, when heated by the sun, works with gravity to move the array to follow the sun. Sometimes, a piston is included in the design to protect the system against shuttering from high winds. They essentially act like a shock absorber.

Both types of tracking systems increase the output of the PV system by 20 to 40 percent.

In addition to categorizing tracking systems as active and passive, you can categorize them as single axis and dual axis. Single-axis trackers follow the sun's azimuth; however, they do not follow its altitude. Dual-axis trackers follow the sun's azimuth and its altitude.

There are some disadvantages to tracking systems:

- Tracking systems require heavy-duty mounts. You usually need pipe that ranges from four to six inches in diameter to set up the tracking system. You'll also require reinforced concrete as its foundation. This increases the cost of the PV system components and installation.
- Tracking systems have additional moving parts, which means more expense and more ways for the system to fail.
- Trackers require continual visual and physical monitoring.
- Trackers require substantially more maintenance than a standard rack-mounted system.
- If the tracking mechanism fails, so does most of the energy production.

You should install the tracking system above the snow line and above any debris that may be below the arrays. Don't forget to account for snow, which can accumulate below the panel as it slides off.

Dual-axis tracking photovoltaic panels at the SunEdison photovoltaic power plant near Alamosa, Colo. This 82-acre tract in south central Colorado, near the New Mexico border, is the site for one of the largest photovoltaic power plants in the United States.
Image courtesy of DOE/NREL

Building-Integrated PV (BIPV)

Building-integrated PV design has many possible applications. With BIPV, the panels and arrays are designed to be part of, and are then built into, the structure. This eliminates the need for standard racks. The panels and arrays shade an area of the structure while still allowing daylight to enter. The arrays can be used as insulation, as a windbreak, or as a sound damper—all while supplying electricity for the structure.

Roofers can use PV panels as roofing materials. The panels replace shingles, tiles, metal sheets, or other roofing materials. Another way to integrate PV panels into a building is to use the panels as awnings or canopy material to provide shade to covered porches or patios. You can also install PV panels in a sawtooth configuration for use as walls or awnings over windows. This allows light to enter the structure, shades the space during the summer, and allows passive heat in during the winter. You can also use PV panels as window and skylight materials. This can be aesthetically pleasing and an interesting feature of the structure.

CHAPTER 5 SUMMARY

This chapter explained the criteria for placing arrays. When placing an array, you must consider possible shading issues, distance from the PCU, aesthetics, maintenance, and systems to keep the PV system safe from vandals.

The chapter also discussed the most popular methods for mounting PV systems. The various mounting options, such as pole mounting and rack mounting, each have advantages and disadvantages. For example, using a pole mount is advantageous when mounting to a structure is difficult or impossible. When you use a tracking system to mount the PV array, the array can follow the sun with minimum maneuvering from the owner. To set up a system that tracks the sun with practically no maneuvering by the owner, you can install a tracking system. BIPV systems are built into the structure; depending how they are integrated, they can provide shading and lighting.

CHAPTER 5 ASSESSMENT

Mounting Systems

1. In North America, modules must face south.
   - ❑ **A.** True
   - ❑ **B.** False

2. Which of the following are vulnerabilities of pole-mounted PV systems? (*Choose two.*)
   - ❑ **A.** Rain
   - ❑ **B.** Animals
   - ❑ **C.** Lower output
   - ❑ **D.** Vandalism

3. Which of the following is an advantage of a pole-mounted system?
   - ❑ **A.** No need for a building
   - ❑ **B.** Safety
   - ❑ **C.** Cooler arrays
   - ❑ **D.** Vandalism

4. Which of the following is a requirement of pole-mounted systems?
   - ❑ **A.** Shade-free location
   - ❑ **B.** Building
   - ❑ **C.** Rocks
   - ❑ **D.** Batteries

5. One reason that designing a system that must be adjusted seasonally is a bad idea is that owners tend to stop doing it after a year or two.
   - ❑ **A.** True
   - ❑ **B.** False

6. With _________________, the panels and arrays are designed to be part of, and then built into, the structure.
   - ❑ **A.** building-integrated PV (BIPV)
   - ❑ **B.** modules
   - ❑ **C.** pole-mounted systems
   - ❑ **D.** rack-mounted systems

7. When installing a pole-mounted system, you must consider terrain and soil type.
   - ❑ **A.** True
   - ❑ **B.** False

8. Tracking systems can increase summer output by up how much?
   - ❑ **A.** 40 percent
   - ❑ **B.** 60 percent
   - ❑ **C.** 50 percent
   - ❑ **D.** 0 percent
   - ❑ **E.** 10 percent

9. What are two advantages of tracking systems? (*Select two.*)
   - ❑ **A.** The owner of the system must manually adjust the PV arrays.
   - ❑ **B.** System output is increased.
   - ❑ **C.** The tracking system is self-adjusting to the sun.
   - ❑ **D.** PV arrays flip over when full of energy.

10. Building-integrated PV panels can do which of the following? (*Select three.*)
   - ❑ **A.** Be used as flooring
   - ❑ **B.** Provide shade for patios and porches
   - ❑ **C.** Be used as a cover for parking spaces
   - ❑ **D.** Need to be washed by hand
   - ❑ **E.** Be used as skylight material

Energy-Storage Device (ESD) System Design

6

THERE ARE MANY FORMS of energy storage. This chapter deals with some of the more common types and provides some initial information on systems that may become more prominent in the future. This chapter looks at energy-storage device (ESD) systems and design. The chapter begins with a survey of battery types and designs. Next, it discusses batteries as an element of the greater PV system. The chapter also covers other energy storage systems, some of which are just coming onto the market. ESD considerations and sizing are a focus of the greater portion of this chapter.

Topics & Concepts

This chapter covers the following topics and concepts:

- PV battery system design
- Batteries in PV systems
- Other Forms of Energy Storage
- Battery sizing
- Sizing a battery bank
- Wiring a battery bank

Goals

After completing this chapter, students will be able to:

- Understand energy storage systems, their strengths and weaknesses, and how they are integrated into PV systems
- Discuss battery design and functions
- List battery types
- Create a battery maintenance schedule
- Review battery subsystem designs
- Explain how to size and wire a battery bank
- List battery safety considerations
- Discuss strengths and weaknesses of energy storage systems
- Discuss the issues that affect storage system life

PV Battery System Design

The battery can be an important part of the PV system. A **battery** stores direct current (DC) energy for use when insolation is low or usage is high. A battery bank is sometimes used with grid-tied PV systems and almost always is used in standalone PV systems.

Batteries lose energy when they are charged. **Charge** is the process of adding electrical energy to a battery. They also lose energy when they are discharged. These energy losses are an important reason for system inefficiencies. Additional panels are sometimes needed when batteries are used in a PV system to offset those losses. Battery maintenance is important to the life of the battery. Keeping the battery in good condition helps maintain the performance of the entire PV system.

Batteries in PV Systems

Battery performance in PV systems depends on many factors. Batteries in PV systems perform in a wide variety of conditions, from very cold to very hot. A number of conditions can shorten the life of batteries. Operating conditions can also affect battery charge, capacity, life, and discharge rates. Extreme temperatures shorten battery life!

It takes time and effort to understand batteries. Batteries have many interesting peculiarities. When you master their somewhat moody characteristics, you can get the most out of them while providing safe, reliable energy for your customer.

All batteries are not created equal. Different types of batteries work in different situations. **FIGURE 6-1** shows an example of a battery cut sheet, which

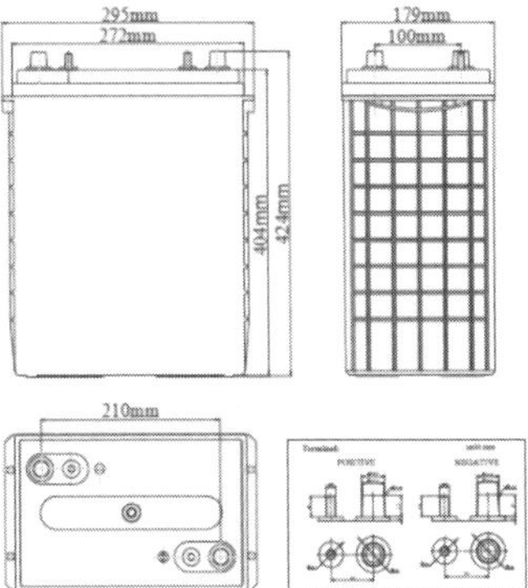

S6-460AGM

Specifications

Nominal Voltage		6V
Rated Capacity (20 hour rate)		415AH
Dimension	Total Height (with terminals)	424mm(16.69inches)
	Height	404mm(15.90inches)
	Length	295mm(11.61inches)
	Width	179mm(7.04inches)
Weight		Approx. 56.0 kg (123.2 lbs)

Characteristics

Capacity 77℉(25℃)	20 hour rate (20A to 5.25Volts)	415AH
	10 hour rate (34.5A to 5.25Volts)	345AH
	5 hour rate (62.2A to 5.1Volts)	311AH
Internal Resistance	Full charged 77℉(25℃)	1.6mΩ
Capacity affected by Temperature (20 hour rate)	104℉(40℃)	102%
	77℉(25℃)	100%
	32℉(0℃)	85%
	5℉(-15℃)	65%
Self-Discharge 77℉(25℃)	Capacity after 3 month storage	91%
	Capacity after 6 month storage	82%
	Capacity after 12 month storage	64%
Standard Terminal	DT	
Max. Discharge Current 77℉(25℃)	2000A (5s)	
Reserve Capacity (Minutes to 5.25V at 80℉(27℃)	@ 25Amps	850Min
	@ 75Amps	240Min
Charging (Constant Voltage)	Cycle — Initial Charging Current 80A Or Small 7.25V~7.45V/77℉(25℃)	
	Float — 6.8V~6.9V/77℉(25℃)	

Discharge characteristics 77℉ (25℃)

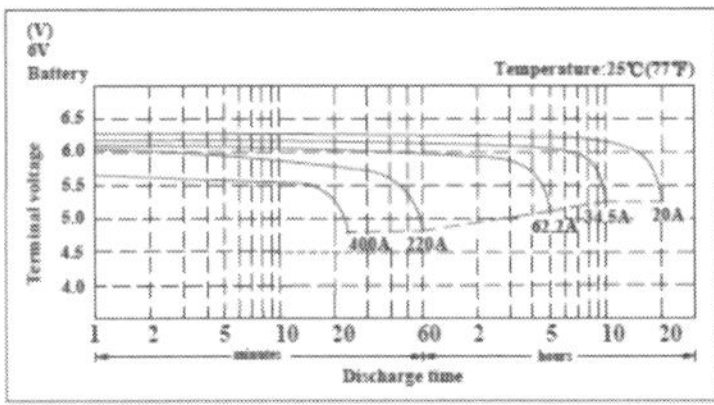

Duration of discharge vs. Discharge current

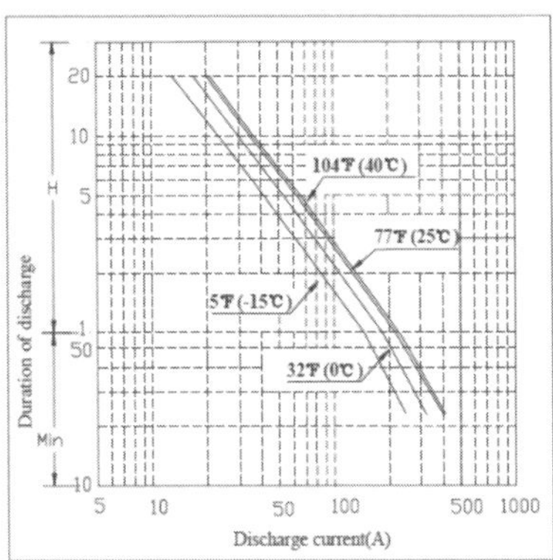

Constant Current Discharge Rating Amperes   @ 77 ℉ (25℃)

Cut off voltage V/cell	30M	45M	1H	2H	3H	5H	8H	10H	12H	20H	24H
1.75V	354	261	212.0	113.0	86.0	60.3	41.3	34.50	29.13	20.00	17.3

FIGURE 6–1 Battery cut sheet.

Courtesy of Surrette

details the characteristics of the particular battery described. It is important to select the correct battery for the PV system. An incorrect battery choice results in shortened battery life and PV system failure. This is an avoidable and very expensive lesson to learn. Proper battery maintenance is important to prolong battery life and get the most out of your batteries.

Batteries for PV systems are tested in controlled laboratories. They are not tested, rated, or certified in real-life PV situations. This makes acquiring real-life data on battery functions difficult. You must learn about batteries from your own experience. Currently, the PV industry does not have specific standards for batteries or battery sizing.

Not all power is immediately used as it is produced by the PV system, so it must be stored. Batteries store unused energy for use when needed. Batteries are also used to stabilize the unstable currents that sometimes occur in almost all electrical systems. Batteries can supply energy when the PV system does not produce enough energy to meet demand.

In reviewing energy-storage devices, it is important to keep the following in mind:

- Energy is required to store energy in the storage device during the charging phase.
- Energy is required to discharge the energy-storage device (ESD).
- Energy is required to hold the ESD at full charge.
- The amount of energy and capacity of charge vary depending upon temperature and the conductor's ability to transfer that energy.
- All these energy requirements are system losses and need to be considered in the use of batteries and any other ESDs.

Battery Design and Construction

Making lead-acid batteries is a toxic and hazardous process. Battery manufacturing involves several processes. Different battery manufacturers use different methods. The final manufacturing stage includes charging and discharging the batteries before they are shipped. For many battery products, the final pre-use charging is completed at one of the following:

- The factory
- The distributor's facility
- The job site

Batteries have the following common components:

- **Battery cell**—The battery cell is the primary electrochemical unit of the battery. It consists of positive and negative plates in an electrolyte solution.

Each cell produces approximately 2.1 volts. The number of cells determines how many volts each battery makes. For example, a six-cell battery will produce approximately 12.6 volts.

- **Reactive material**—These materials are located in the cell and compose the positive and negative plates that react to the battery's electrolyte and specific chemistry. The amount of material in the plates determines the capacity of the battery. For example, lead-acid batteries contain lead dioxide (positive plate) and a metallic sponge lead (negative plate) that react with sulfuric acid.
- **Electrolyte**—The electrolyte is a substance containing free ions that makes the battery's particular chemistry electrically conductive. This causes the ionic transfer between the plates.
- **Grid**—A grid is a framework usually made from a lead alloy in lead-acid batteries. It conducts current and supports the reactive material on the plate. Calcium and antimony are commonly used to make the grid stronger. The grid can affect the performance of the battery. Some grids are made into a web. Other grids are made by tubular plates.
- **Plate**—The plate is composed of the grid and reactive material, and is known as the electrode. Each cell has several positive and negative plates. The mass of the plate determines how deeply the battery can cycle. The best type of battery for forklifts, electric cars, and PV systems are deep-cycling batteries.
- **Separator**—A separator is a porous insulator that divides the negative and positive plates. The separator allows the ions in the electrolyte to flow to and from the negative and positive plates. It also keeps the plates from touching each other and shorting out the battery. Separators are typically made from rubber, glass wool, or plastic.
- **Element**—The element is the assembly of the plates stacked or strapped together.
- **Terminal posts or terminals**—Terminal posts are located outside of the battery. They are the positive and negative connections to the battery. Terminals should be periodically inspected, tightened (not over-tightened), and cleaned.
- **Vents**—Gases are produced when most lead-acid batteries are charged. These gases pass through a charcoal filter, and are then released to the atmosphere through a vent in the battery.
- **Case**—The battery's case is usually plastic, glass, or hard rubber. It holds the electrolyte, the plates, and the separators. The case is usually enclosed. Cases are sometimes transparent to simplify monitoring. This makes it easy to see the condition of the plates and the electrolyte level.

Battery Types and Classifications

There are two main types of batteries: primary and secondary. There are also many classifications of batteries. This variety enables designers to design batteries for particular uses. For example, lead-acid batteries are often used in PV systems. Nickel-iron or nickel-cadmium batteries are good choices if the PV system is located where temperatures are low.

Battery Types

Non-rechargeable batteries are called primary batteries. Primary batteries are used in electronic devices such as flashlights and radios. They are not suitable for use in PV systems because they are not rechargeable.

Rechargeable batteries are called secondary batteries. They are used in cars and trucks, and are usually lead-acid batteries. The battery is recharged when the external electrical potential difference between the terminals is greater than the internal electrical potential difference, thereby reversing current flow.

Lead-Acid Battery Classifications

A lead-acid battery has three main classifications. The best battery classification for use in PV systems is the traction, or motive power, lead-acid battery. **FIGURE 6-2**

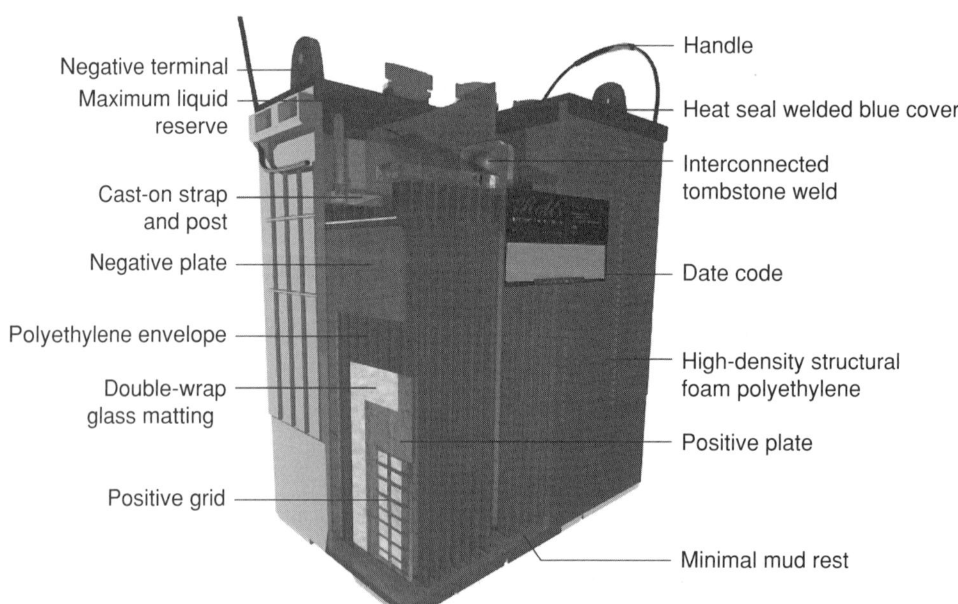

FIGURE 6-2 Lead-acid battery internal components.

shows the internal components of a lead-acid battery. Batteries in this classification are characterized as follows:

- They have very thick plates.
- They are heavy-duty.
- They have deep-cycle capability and characteristics.
- They can be discharged and recharged deeply many times.

The second classification of lead-acid battery is the starting-lighting-ignition (SLI) battery. SLI batteries are characterized as follows:

- They are used in PV systems installed in developing countries.
- They work well in very small standalone PV systems.
- Their average daily depth of discharge (DOD) should not exceed 20 percent.
- Their maximum allowable depth of discharge should not exceed 60 percent.
- They have a life expectancy of about two years.

A third classification of battery is the stationary battery. These batteries are used for critical load devices such as telephones and computers. They can be deeply discharged on occasion and do not permit as much cycling. Stationary lead-acid batteries are not suitable for PV systems.

Types of Lead-Acid Batteries

There are two main types of lead-acid batteries used in PV systems: liquid vented and sealed. A **sealed battery** is also called a valve regulated lead-acid (VRLA) battery. When VRLA batteries are overcharged, the pressure rises and the cell vents open. VRLA batteries do not tolerate being overcharged because the electrolyte cannot be replenished, and repeated overcharging will kill the battery.

Lead-Acid Battery Chemistry

Lead-acid cells have positive plates made of lead dioxide and negative plates made of sponge lead. The electrolyte is usually a solution of diluted sulfuric acid. The battery produces a current when the lead dioxide and the sponge lead react with the sulfuric acid and convert to lead sulfate.

Absorbed Glass Mat Batteries

Absorbed glass mat (AGM) batteries are a type of lead-acid battery. They tend to be of the VRLA variety. That means most of them are sealed batteries. The glass mats absorb and hold the electrolyte in place. The mats are located between the plates.

You should not allow AGM batteries to overcharge. AGM batteries are intolerant of high temperatures. When AGM batteries are almost fully charged, they give off hydrogen and oxygen gases, which escape through vents. You should periodically add water to the battery if the battery type allows it. It is best to set and monitor charge settings to prevent venting of gases and water by carefully controlling the charging of the AGM battery.

Gel Cell Batteries

Gel cell batteries are similar to AGM batteries, except the electrolyte is held in place by a gel solution. You can install these at any orientation.

Valve Regulated Lead-Acid (VRLA) Batteries

There is no mystery to VRLA batteries. There are, however, a few things to remember.

VRLA batteries are normally used in PV systems because of owners' general lack of enthusiasm for battery maintenance. In fact, valve-regulated batteries are often called *maintenance free.* This applies only to watering the battery. Removing the burden of watering does result in less maintenance, but it doesn't make the battery maintenance free! All other battery maintenance is still required if you wish the batteries to have a full and productive life. Lead-acid batteries can be destroyed in one season due to lack of attention to charge settings and watering.

Although VRLA batteries tend to have shorter lives than other varieties, they should be used if you do not believe the owner will consistently water and check the batteries monthly for the life of the system.

A good rule of thumb is to test hot electrical connections three times to make sure of the following:

- The voltage is right.
- The polarity is correct.
- The value settings on the meter are set accurately.

It may save your life, help you avoid a shock, or prevent a fire.

Nickel-Cadmium Batteries

A **nickel-cadmium (NiCd) battery** is a secondary type of battery, and so rechargeable. They have many advantages over lead-acid batteries. These advantages make them ideal for use in standalone PV systems.

The advantages of nickel-cadmium batteries are as follows:

- Their loss of capacity at −20 degrees Celsius is less than 20 percent, while lead-acid batteries lose more than 40 percent of capacity at such temperatures.
- The number of charge/discharge cycles is approximately twice as many as that of lead-acid batteries.
- They may be stored in a fully discharged state without any negative side effects.

- They have tolerance of overcharging for indefinite time periods.
- They offer stable output voltage as the battery progresses through its discharge cycle. In contrast, the voltage of a lead-acid battery decreases linearly and much more rapidly.
- They are relatively easy to maintain because they are sealed and so do not need to have the electrolyte replenished.

Nickel-cadmium batteries also have few disadvantages:

- The nominal voltage is 1.2 volts. Unlike a lead-acid battery, which requires only six cells, a nickel-cadmium battery needs 10 cells to create a 12-volt battery.
- A nickel-cadmium battery's voltage radically drops when it is almost completely discharged.
- Nickel-cadmium batteries are more expensive than lead-acid batteries.
- Nickel-cadmium batteries are not always available and require more advanced planning to incorporate into a project.
- Unlike lead-acid batteries, which maintain approximately the same amount of capacity as the battery ages, the capacity of nickel-cadmium batteries linearly decreases.

There are two types of nickel-cadmium batteries: pocket plate and sintered plate. You can add water to pocket-plate nickel-cadmium batteries in very rare cases. However, nickel-cadmium batteries are mostly sealed batteries. They are used in remote locations and use an alkaline potassium hydroxide solution. These batteries are expensive to buy, but they have lower life-cycle costs than most other battery choices.

Sintered-plate nickel-cadmium batteries have a metallic case. The plates are immobilized. This prevents leaking. The battery can be oriented in any direction when installed because it is sealed. The main disadvantage to using a sintered-plate nickel-cadmium battery is that after discharging and recharging, the battery memorizes the pattern. This memory limits the recharge rate and varies with battery quality. This eventually does not fully recharge the battery. You must perform a special charge and discharge routine to restore the initial capacity of the battery.

Nickel-Iron Batteries

The nickel-iron (NiFe) battery is a deep-cycle storage battery with a chemistry of nickel (III) oxide-hydroxide cathode and an iron anode. The electrolyte is usually potassium hydroxide. Nickel-iron batteries are some of the oldest used. They were quite common in the early 1900s for use in DC energy storage and electric cars. They were invented in 1899 by Swedish inventor Waldemar Jungner.

Thomas Alva Edison improved the design and performance of these batteries in 1901.

Nickel-iron batteries are very durable. They can take a lot of abuse through overcharge, over-discharge, and limited short-circuiting. They can have a very long life—20 to 40 years, if properly maintained. Nickel-iron batteries do not produce sulfate (even when left discharged for years at a time), are less toxic, and are more resistant to cold temperatures. They also do not need frequent equalization charges.

Nickel-iron batteries can be very deeply discharged with little reduction in lifetime cycling. They can survive some plate exposure. They do not produce toxic outgases, unlike lead-acid batteries. With nickel-iron batteries, you can reduce the number of batteries required to meet your power goals due to the debt of discharge (DOD) capability. DOD can be as high as 80 percent but may be better if set to 50 percent. Occasional deep discharge is good for most batteries.

On the negative side, nickel-iron batteries are heavier and not as efficient in charging and discharging. They therefore require more energy to meet their full capacity than a lead-acid battery with a coulombic efficiency of 65 percent. The standby discharge rates can vary from 20 to 30 percent per month, which eliminates them from the seasonal charge capability.

These batteries do not use acid. They use a base of potassium hydroxide (KOH) and lithium hydroxide combined with distilled or deionized water. Water constitutes approximately 85 percent of the electrolyte mix and requires different safety procedures than lead acid.

Lithium-Ion Batteries

Lithium-ion batteries (LIBs) use metallic lithium, which is highly reactive. Because some types are more susceptible to thermal runaway, there may be safety hazards. Explosions and fire are possible hazards when lithium batteries are over-charged or over-discharged. Fires can also start if high temperatures, over-current conditions, or a short circuit occurs.

LIBs are growing in popularity for military, electric vehicle, and aerospace applications. Lithium-ion batteries are used in cell phones, computers, portable appliances, cameras, electric cars, and new PV products that are coming to the market. They have a nominal cell voltage of 3.6 volts. Lithium-ion batteries may not always be a good choice for use in PV systems, however, because they are still expensive.

Following are the advantages of using lithium-ion batteries:

- They have high open-circuit voltage.
- They come in many shapes and sizes.

- They have no memory effect.
- They are lighter than most energy density–equivalent secondary batteries.
- They have low (5 to 10 percent per month) self-discharge rates.

Following are the disadvantages of using lithium-ion batteries:

- Capacity loss takes place with high temperatures and high charge levels, which means it may not be best to fully charge batteries in some instances.
- Cell capacity diminishes due to charging deposits, which results in an internal lowering of the battery's ability to deliver current.

Air-cooled lithium ion battery module. Lithium ion batteries have high power and energy.

Courtesy of DOE/NREL

Other Forms of Energy Storage

Batteries are not the only form of energy storage available, however. Two other intriguing technologies are hydrolizers and flywheels.

Hydrolyzers

One of the more tempting technologies that can affect the on- and off-grid world is using PV to break water into two high-quality elements and fuels. Hydrolysis is the electrochemical process by which electricity is used to break water into its component parts of hydrogen and oxygen.

PV is a natural for hydrolysis when there is excess energy available, which would normally be wasted. Currently, hydrolyzers are in use in some large installations. The future of hydrolysis and PV, however, will be in the development of smaller, more cost-effective units where every kWh of energy can be used. Hydrogen can be compressed and used later to power a fuel cell or an internal-combustion engine. As a result, a properly sized PV system could supply all the energy needed both on and off the grid.

When water is hydrolyzed, it is self-compressing at a very small internal energy loss, which may result in the lack of need for a high-pressure compressor pump.

Flywheel Energy Storage (FES)

Flywheel energy storage is like a kinetic battery. It works by accelerating a rotor (flywheel) to very high speeds (20,000 to more than 50,000 rpm or higher in a vacuum enclosure). This maintains the energy in the system as rotational energy.

When energy is removed from the system, rotational speed is reduced as a consequence of the principle of conservation of energy. It is a flexible storage medium for short- and medium-term electrical storage, where adding energy to the system results in an increase in the speed of the flywheel.

Flywheels can be brought up to speed in a few minutes. This is more rapidly than many other forms of energy storage. As in many applications, the flywheel is in continuous motion. When electrical energy is needed, the spinning drum provides it by driving a generator.

Modern high-energy flywheels being developed use composite rotors made of carbon-fiber materials. They have very high strength-to-density ratios with rotation speeds of up to 100,000 rpm in a near vacuum to minimize aerodynamic losses. The use of superconducting electromagnetic bearings may almost eliminate friction. This could extend their cycle life and reduce maintenance.

Battery Performance Characteristics

In order to choose the appropriate battery for a PV system, it is important to understand how batteries work. The following are terminology and definitions to help you understand batteries:

- **Ampere-hour (Ah)**—This is the unit by which a battery's electrical storage capacity is measured. It is called ampere-hours, or amp-hours, because you multiply the discharge current in amperes measured over a specific period of time (hours). One ampere-hour equals the transfer of 1 ampere over one hour, which equals 3,600 coulombs of charge. If a battery delivers 5 amps over a 20-hour period, the battery has delivered 100 ampere-hours. Battery ampere-hour ratings depend on the rate at which they are discharged at a constant current. A more rapid discharge results in less available energy produced from the same amount of battery capacity. A battery may include a chart showing discharge rates of 1 to 120 hours. The battery nameplate will have a specific rating for a specific rate of discharge at constant current. In PV design, systems are calculated at the 20-hour rate due to the mixed rate of discharge during normal operation.
- **Battery capacity**—This measures how much electrical energy a battery can store and deliver. Capacity is generally discussed as a rate of discharge over a period called ampere-hours. Capacity depends on the quantity of active material, the design of the plates, and the electrolyte. Operational factors affecting capacity include the rate of discharge, the discharge depth, the voltage cutoff, the age of the batteries, the charging cycle, and the internal battery temperature. Historically, battery capacity is quoted in ampere-hours (Ahs), but can also be discussed in kilowatt-hours (kWhs) with a simple conversion.

- **Cascading battery failure**—This refers to the imbalances that occur between cells and batteries when batteries of different ages or usage levels are combined in a battery array. This may begin with a bad cell. The result is a much higher resistance (which accounts for a high temperature rise) than is allowed by the design. Cascading battery failure begins with the weakest battery in the pack. It may be due to a number of factors that result in that initial battery destroying the next one or two. The effect then cascades until the batteries are all dead. The cascading effect takes place in a number of electrical and technical components. With batteries, this can result in damage and potentially explosions. Never add new batteries to older batteries, as the new batteries can accelerate the process and can begin this chain reaction.

- **Cutoff voltage**—The lowest discharge voltage set by the battery manufacturer.

- **Cycle**—Cycle refers to the discharge and recharging of the battery. Cycling can be confusing because the number of cycles is related to the rate of discharge, the depth of discharge (DOD), voltage cutoff, age, charging cycle, number of cycles, and temperature. A battery may last one cycle or 5,000 or more, depending upon its type and other factors.

- **Depth of discharge (DOD)**—This refers to the amount of capacity withdrawn from the battery compared to the battery's full capacity expressed as a percent.

- **Discharge**—Discharge occurs when a battery delivers current to a load.

- **Charge**—Charge occurs when a battery receives current to store.

- **Rate of charge/discharge**—Expressed in hours, the rate of charge/discharge is the ratio of the nominal battery capacity to the charge or discharge.

> **NOTE**
>
> When designing battery arrays, pay special attention to the consistent size and length of cables, the balance of the battery strings, the distances between batteries, the strings, and the load. Allow for sufficient airflow and expansion space between batteries. Improper and inconsistent cable sizes can damage battery arrays.

This photo shows a battery array with cables in disarray. The dilemma is to shorten the battery cable lengths for balance and less resistance or neatly install the cables for improved safety. What would you do?

Courtesy of Outback Power

GETTING THE MOST OUT OF YOUR BATTERY

Follow these tips to get the most out of a battery backup system:

- Make sure all the required battery maintenance is done monthly.
- Do not mix new batteries with old.
- Try to keep the operating temperature as close to 25 degrees Celsius as possible.
- Do not discharge lead-acid batteries more than 20 percent except in case of emergency or as part of the maintenance plan.
- Ensure you select the charge control for the correct battery charge characteristics and profile for the battery type being used.
- Use a temperature-compensation sensor.

- **Negative (–)**—When a battery is discharging, electrons flow from the lower electrical potential point—that is, from the negative battery terminal.
- **Positive (+)**—When a battery is charging, direct current flows from the higher electrical potential point, the positive battery terminal.
- **Open-circuit voltage**—This describes the battery's state when it is at rest, not charging or discharging. It is the electrical potential difference between the terminals.
- **State of charge (SOC)**—Like a gas gauge in a car, the state of charge (SOC) is the amount (state) of charge from 0 to 100 percent compared to the maximum capacity of the battery.

Batteries in standalone PV systems are charged differently from batteries charged at the factory. There are five possible charging stages:

- **Bulk or boost stage**—The majority of the charge comes during the bulk charge, where high current and voltage fill the battery to a capacity between 70 and 90 percent of capacity. You can perform bulk-charging at any rate as long as the bulk-charging does not cause the voltage to exceed the gas voltage limit.
- **Absorption stage**—High voltage is maintained while the current is reduced until the battery is charged to its full capacity.
- **Float stage**—When a battery is essentially full, a float or finishing charge occurs. A small trickle or pulse charge allows the battery to hold an optimal voltage. This limits the amount of overcharge or undercharge to the battery.

- **Equalization stage**—This maintenance process prevents battery stratification and reduces sulfates by raising the battery voltage to allow the electrolyte to bubble, or "boil." This refreshes the cells and brings them to a more internally equal state to lengthen the battery life and increase the battery's performance. An equalizing charge is used to maintain consistency among cells.
- **Maintenance stage**—This is where the battery is continually monitored and automatically topped off as required. Also called the sleeper or snooze stage.

As mentioned, the depth of discharge (DOD) is the percentage of the battery's capacity that is withdrawn from the battery compared to the battery's total capacity. There is an allowable maximum DOD and the average daily DOD. Both will affect the life of the battery.

The allowable daily DOD depends upon the maximum of the capacity that can be withdrawn from a battery. The cut-off voltage and the discharge rate control the DOD. The allowable DOD is also controlled by low insolation, excessive use, and low temperatures.

A **deep-cycle battery** allows for a 20 percent to as low as 75 percent discharge. Note, however, that a 50 percent DOD erodes battery cycle life. **Battery cycle life** refers to the number of complete charge/discharge cycles a battery can perform before its nominal capacity falls below 80 percent of its initial rated capacity.

Autonomy in relation to the capacity the system requires during periods of low insolation depends on the DOD of the batteries. Batteries with a lower DOD will create systems with shorter periods of autonomy.

Average daily loads often vary by season. In the winter, average daily loads usually increase, and the battery capacity decreases. If the PV system is designed for substantial autonomy, the DOD may be a small percentage of the battery capacity. The DOD will be higher if the PV system is not designed with a large battery bank.

The percentage of energy capacity in a battery is referred to as its state of charge (SOC). When a battery is discharged, the battery is said to have a decrease in SOC. Charging a battery increases its SOC. For example, a battery that is 25 percent discharged has an SOC of 75 percent.

The time a completely charged battery is able to supply energy by itself (independently of energy input) for the daily load is called the autonomy period. During the autonomy period, little or no energy is supplied by the PV array. An autonomy period may last anywhere from an hour to six days or longer. In many instances, the longer the autonomy period, the lower the average daily DOD, and the lower the probability that the allowable DOD, or minimum load voltage, will be reached. The alternative to a high DOD is to add batteries or use a generator to maintain power for the loads.

Batteries have self-discharge rates. **Battery self-discharge** occurs when there is an open-circuit mode without any charge or discharge current. Discharge rates vary by battery type. The active ingredients and lead alloy used in the cell grid affect the self-discharge rate. High temperatures have an adverse affect on the self-discharge rate. In general, for every 10 degree Celsius increase in temperature, the self-discharge rate doubles.

The lifetime of a battery depends on the following:

- PV design
- Battery components
- Frequency and depth of discharges
- Charging frequency and state
- Charging methods
- Temperature

Proper battery care is essential to battery life. Batteries should be kept at moderate temperatures as close to 25 degrees C as possible. Do not overcharge or over discharge batteries. The lifetime of the battery is directly related to the quantity and depth of discharges.

Temperature affects battery performance and life. Battery performance changes dramatically with temperature. At higher temperatures, battery life decreases, but performance and capacity increase. Chemical reactions accelerate at higher temperatures. Similar to the self-discharge rate, for every 10 degree Celsius increase in temperature, the reaction rate doubles. This creates gassing and loss of electrolytes. But the opposite also occurs with lower temperatures. Because lower temperatures slow chemical reaction rates, battery life increases, but performance and capacity decrease.

Discharge rates affect the capacity of the battery. Battery capacity lowers when the discharge rate is high. This is important in design when you look at battery capacity on the label. The rate of discharge affects the usable amount of energy available, although the battery may be exactly the same physically and electrically. So a 5 ampere-hour rating provides less energy than a 100 ampere-hour rating. A 20-hour rating is commonly used in PV system design.

Electrochemical activity causes corrosion. Electrochemical activity is the result of immersing two dissimilar metals in an electrolyte and affecting them with a current in charge or discharge mode. Corrosion also occurs when two dissimilar metals come in contact with each other. This causes a reaction called a short. Electrolyte gassing at the battery terminals causes corrosion. Higher temperatures accelerate corrosion. Corrosion shortens battery life.

Gassing happens when a battery is charging and is almost fully charged. Cell voltage rises dramatically at this point. If the battery is sealed, the gas stays in the cell until it reaches a predetermined pressure. If exceeded, the cell is vented, and the gas and water are released. Gassing uses part of the charge current and reduces the battery's efficiency.

Gassing is necessary during equalization in flooded batteries. The equalization charge allows gassing to agitate the electrolyte in the battery and prevents stratification. It also knocks potentially loose lead sulfate crystals from the plates. This keeps them from bridging the gap between positive and negative, therefore discharging and shorting the battery. Battery performance declines when electrolytes are not mixed. Battery life decreases if over-gassing occurs.

The maximum voltage a charge controller allows a battery to achieve also controls gassing. The risk for gassing increases with the voltage the batteries reach. Charge controllers allow high rates of charging up to the gassing point under normal circumstances. When the gassing point is reached, the charge controller disconnects or dramatically reduces the PV current. Disconnecting prevents overcharging.

The exception to this is during the planned process of providing an equalization change. Equalization changes are for battery maintenance.

Gassing is caused by other factors as well. An increase in temperature can cause an increase in gassing. Temperature and gassing effects determine which controller should be chosen.

Battery System Design and Selection Criteria

There are many battery types to choose from when selecting the battery type for a PV system. When choosing a battery, consider the PV system's application. Reviewing manufacturers' literature and gaining field experience helps narrow the choices.

Use the following list of considerations to help narrow the choices of batteries:

- System type and mode of operation
- Characteristics of charging, such as charging profile
- Autonomy issues (required hours or days of energy storage)
- Amount of discharge current needed and its variability
- Allowable maximum discharge depth
- Daily depth of discharge requirements
- Accessibility of location
- Environmental conditions, including humidity, cold and hot temperatures
- Cyclic life calculated in years (1,000 cycles is about 2.7 to 3 years, depending upon usage)

- Maintenance requirements
- Unsealed or sealed
- Self-discharge rate
- Maximum cell capacity
- Energy-storage density
- Weight and size
- Characteristics of gassing
- Freezing susceptibility
- Susceptibility to sulfation
- Concentration of the electrolyte
- Auxiliary hardware availability
- Terminal configuration
- The manufacturer's reputation
- Warranty
- Cost and expectation for the useful life of a battery

Battery Subsystem Design

After selecting the appropriate batteries, the PV designer decides how to configure the batteries in the PV system (see **FIGURES 6-3**, **6-4**, and **6-5**). The designer must make decisions about the number of batteries required and how they will be installed: in parallel or in series. The designer must also select battery cables, terminals, and sizes.

In series installation, the negative terminal is connected to the positive terminal of the next battery. This sequence repeats until all the batteries are connected to meet the appropriate voltage requirements. The total voltage of batteries in series is the sum of the battery voltages. That being said, current ampere-hour (Ah) capacity for the series is the same capacity as it is for one battery. If batteries of different capacities are wired together, the capacity of the series is the capacity of the lowest battery capacity and voltage. This is unwise and risks the life and safety of the array.

Current has more than one path to follow when batteries are installed in parallel.

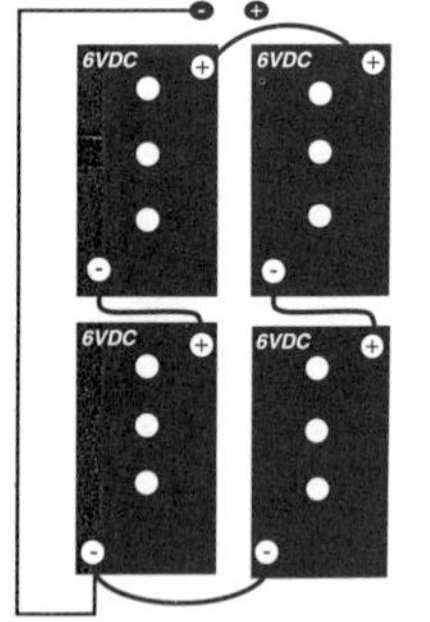

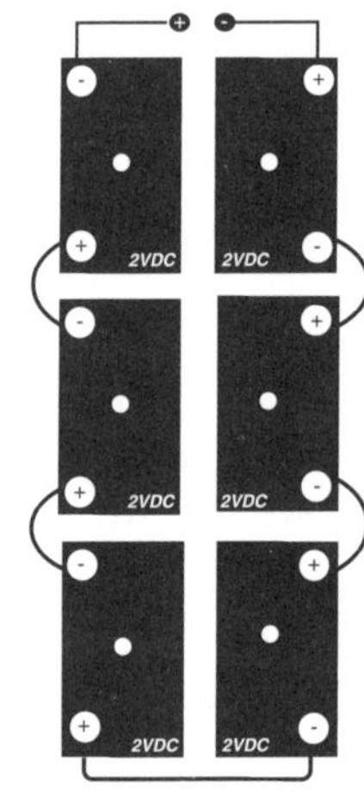

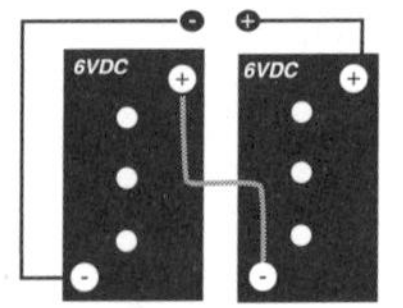

FIGURE 6-3 12-V battery configurations.

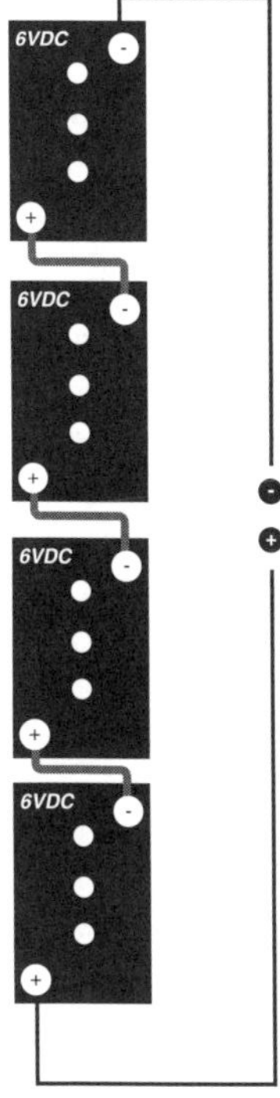

FIGURE 6-4 48-V battery configuration.

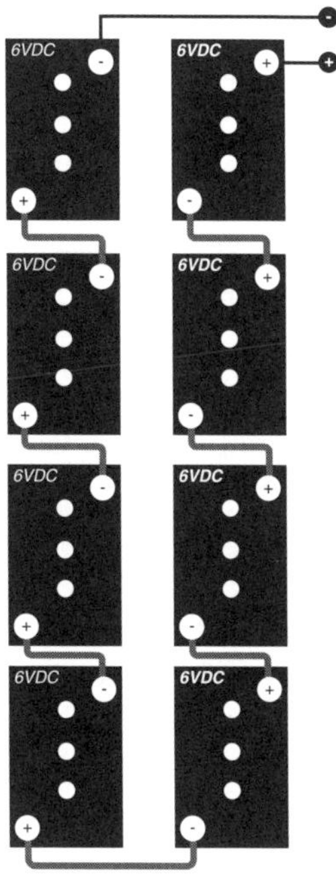

FIGURE 6-5 24-V battery configuration.

In parallel, the positive terminals are connected together, as are the negative terminals. The voltage is the same across the circuits. The capacity is the sum of the parallel branch capacities.

Battery manufacturers prefer battery systems to have few parallel strings and a limited numbers of batteries. Differences in voltages occur when there are many parallel strings. This voltage imbalance causes the capacity in some series to become unequal. This puts strain on the circuit, with the highest resistance causing the other batteries to not charge at the same rate. This overworks some batteries in the battery bank and underworks others. Charging batteries at different rates shortens battery life.

Load voltage requirements dictate the selection of the battery bank voltage. Most remote standalone PV systems require 12, 24, or 48 volts, with 48 volts being the most common. Use higher volt-rated battery arrays to lower the current and resistance for PV system. Using lower currents cuts costs by requiring the use of smaller conductors, fuses, and disconnects.

An improperly sized conductor results in poor battery and array performance. So does a poor battery terminal connection. Conductors are used to connect the battery to the other components of the PV system. The current requirements dictate the required conductor size.

You must also consider voltage drop and environmental conditions. You should select conductors to handle at least 125 percent of maximum current. They should also limit voltage drops to less than 5 percent for moderate performance. For high-performance systems, the voltage drop must be less than 2 percent. Hot climates demand even less than that. Shortening the cable runs and/or increasing the conductor size will accomplish this. Insulating materials for conductors should be based upon resistance, moisture and temperature. Size the conductors to handle very high currents flowing between the battery and the inverter.

In planning your array, design for the shortest-yet-equal lengths of cable between batteries. Do the same between the different strings of batteries. Also determine what the manufacturer's torque requirements are for terminal connections and carefully torque to suggested requirements.

You must install overcurrent and disconnect devices with batteries. Overcurrent devices reduce the probability of short circuits, explosions, burns, fires, electrocution, and equipment damage. Both positive and negative conductors must have disconnects if the system is not grounded. If the system is grounded, disconnects are necessary only for the ungrounded conductor.

Battery Auxiliary Equipment

The location of the batteries has a direct effect on their performance. Local electrical codes require safety to be designed into the battery location. Batteries must be in an enclosed area away from controls and other PV system components. Batteries must not be in areas where any metallic objects can be dropped on them. The enclosure may need to be insulated, climate controlled, and adequately vented.

Wide variations in temperature affect battery performance. A passive thermodynamic enclosure is best for battery performance. These types of enclosures do not require equipment that uses energy, such as fans or air conditioners.

Ventilation is also an important safety consideration. Many batteries produce hydrogen gas and other explosive and toxic fumes that must be vented away from the battery enclosure. Vents or ducts may be adequate, however, fans may be required to provide enough air exchange to minimize gas buildup. Personnel should not be allowed in battery storage areas unless they are performing maintenance tasks. Keep batteries in a locked area for safety.

Provide battery containment in the event of a crack or spill from the battery bank. The spill containment must hold all of the electrolyte in the batteries, whether flooded, AGM, or gel battery type.

Another aspect of batteries is monitoring. Monitoring can be simple or complex. Simple systems use voltage and current meters or battery charge indicators. Readings are taken manually, and recorded. A complex system uses automated devices to record current, voltage, temperature, and water levels. Batteries in small PV systems are monitored as part of routine maintenance checks.

Battery Maintenance

Battery maintenance varies widely depending on design and application. Battery manufacturers provide recommended maintenance tasks. Some batteries require more maintenance than others. All batteries require some monitoring and maintenance.

Maintenance tasks may include adding water, retightening terminals, cleaning terminals, battery cases, and cables, and checking performance.

To ensure batteries operate at full capacity, you should monitor (and keep a log of) the following:

- Specific gravity recordings of electrolyte
- Conductance readings
- Temperatures
- Cell voltage
- Results of capacity tests
- Voltage and current readings while batteries charge

Battery Maintenance Requirements

Flooded batteries need water. If needed, add water after fully charging the battery. Doing so prevents overflow while charging.

Check flooded batteries once a month until you determine their watering needs. Watering needs depend on the local climate, charging methods, application, etc. Follow the monthly maintenance practice with a printed schedule. Take the readings and inspect everything to get the longest life and highest performance possible out of your batteries.

Here are some important things to remember:

- Never expose the plates to air. This corrodes the plates.
- The electrolyte level should be below the vents, about half to three-quarters of an inch above the plates.
- Use distilled or deionized water only.

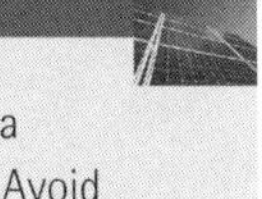

NOTE

Caution: The electrolyte is a solution of acid and water. Avoid contact with skin!

Here is the watering procedure, step by step:

1. Open the vent caps and look inside the fill wells.
2. Check the electrolyte level. The minimum level is at the top of the plates.

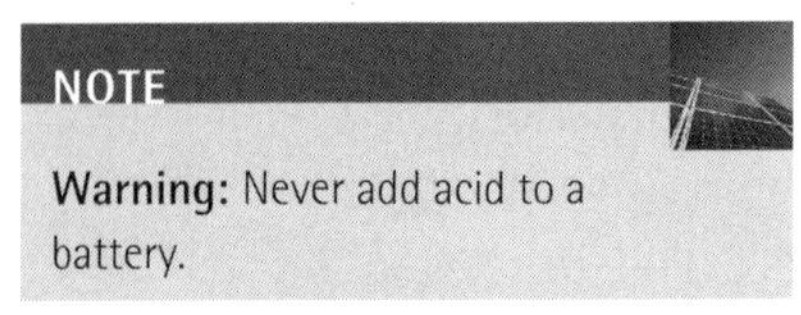

Warning: Never add acid to a battery.

3. If necessary, add just enough water to cover the plates.
4. Put the batteries on a complete charge before adding any additional water.
5. Once charging is complete, open the vent caps and look inside the fill wells.
6. Add water until the electrolyte level is 1/8-inch below the bottom of the fill well. You can safely use a piece of rubber as a dipstick to help determine this level.
7. Clean and replace all vent caps.

Battery Safety Considerations

Batteries use hazardous materials and chemicals to store energy. The energy stored in the batteries can also be dangerous. In short-circuit conditions; batteries have the capacity to deliver several tens of thousands of amps in a fraction of a second.

You must follow safety procedures when storing, using, and performing maintenance tasks on batteries. Wear personal protective equipment (PPE) when working around batteries. Do not wear jewelry. Appropriate PPE includes the following:

- A face shield
- A ventilation mask
- A leather, rubber, or plastic apron
- Gloves

Use caution when handling the battery's electrolytes, whether acid or base. If electrolyte splashes into your eyes, force the eyes open and rinse with clean, cool water for 15 minutes. If you ingest electrolyte, immediately do the following:

1. Drink water or milk, followed by vegetable oil, raw beaten eggs, or milk of magnesia.
2. Call your local poison control center.

If sulfuric acid gets on your skin or clothing, use a solution of baking soda, and water to neutralize and remove it.

Follow safety procedures when mixing water with the electrolyte. Pour the acid into the water slowly. Use nonmetallic utensils such as containers and funnels to mix the electrolyte and water.

Electrolytes like potassium hydroxide (KOH) are highly basic. They form strong alkaline solutions in water. The material safety data sheet (MSDS) for potassium hydroxide includes the following:

- **Eyes**—Immediately flush eyes with plenty of water for at least 15 minutes, occasionally lifting the upper and lower eyelids, then get medical aid.

- **Skin**—Flush skin with plenty of soap and water right away for at least 15 minutes while removing contaminated clothing and shoes. Get medical aid. Discard contaminated clothing in a manner that limits further exposure of anyone or anything to the chemical.
- **Ingestion**—Do NOT induce vomiting. If the victim is conscious and alert, give him or her two to four cupfuls of milk or water. Never give anything by mouth to an unconscious person. Get medical aid immediately.
- **Inhalation**—Get the person away from the gas and out into fresh air immediately. If the person's breathing is difficult, provide oxygen. If the person has stopped breathing, apply artificial respiration. Get medical aid immediately.
- **Notes to physician**—Treat symptomatically and supportively. Medical attention should be sought immediately if the electrolyte comes in contact with the eyes or skin, or is ingested or inhaled.

Other safety considerations are as follows:

- Install eyewashes and safety showers.
- Keep a fire extinguisher nearby.
- Be aware that fumes from batteries can be a problem.
- Locate batteries in an area with adequate ventilation to avoid buildup of explosive gases.
- Use battery vent caps and a flame arrester to avoid explosion.
- Keep batteries appropriately charged.
- Keep all ignition sources such as cigarettes, flames, and other sparks away from batteries.
- Make sure there is no load on the batteries when connecting or disconnecting to keep arcing or sparks from occurring.

Batteries are hazardous materials and cannot be disposed of in landfills. Following are several ways to dispose of batteries properly:

- Some manufacturers and distributors provide disposal sites for batteries.
- Landfills often have battery-recycling programs. Batteries can be taken to the landfill and placed in battery recycling containers. From there, the batteries are transferred to legal battery-recycling centers.

Battery Sizing

Battery sizing enables the battery system to meet the load demand with no contribution from the photovoltaic system. The principal goal of battery storage in a standalone PV system is to ensure that the annual minimum photovoltaic system energy output equals the annual maximum load requirement for energy input or equals it with some integrated backup. The photovoltaic system must also maintain a continuous energy supply at night and on cloudy days. The amount of battery storage needed depends on the load energy demand and weather patterns at the site. Having too much energy and storage capacity increases the system cost, therefore, there is a trade-off between keeping the cost low and meeting the energy demand during low solar energy periods.

Some design possibilities include the following:

- Using the surplus solar energy to perform various tasks that can be switched on when excess energy is available.
- Undersizing the storage capacity and array and meeting periods of deficiency with auxiliary power such as a generator. For greatest reduction in fuel costs, take manual control of the generator. Make sure it has a low-voltage alarm. It does little good to run the generator to fill the battery late at night or early in the morning if the PV system can do it after the sun rises.
- The load may have to diminish or stop its energy demand during brief periods when insolation is low.
- Learn to do with less energy and, in a few instances, do without.

One of the instructions often given new off-grid PV system owners is this: Don't panic! You are going to run out of electricity three times before you get the hang of living off-grid. In time, you will become much more aware of the daily and seasonal flows of energy and climate, and you will enjoy the fact that there is no reason to panic if the power goes out. It is temporary!

Depth of Discharge

As mentioned, the depth of discharge is the percentage of rated battery capacity energy that is withdrawn from the battery. The capability of a battery to withstand discharge depends on its construction and the amount of time the energy is drawn at a constant current. If you used 20 percent of the battery energy, the DOD is 20 percent.

Temperature Correction

Battery performance increases with temperature. At higher temperatures, batteries store more energy; however, battery life is shortened. The battery can store less energy at colder temperatures. The battery may freeze in extremely cold temperatures. Temperature correction is needed to adjust the temperature to which the battery will be subjected during its operation and the discharge rate expected. This can be accomplished with insulation, an incandescent light, or a small space heater. For better battery charging, use a remote battery temperature sensor.

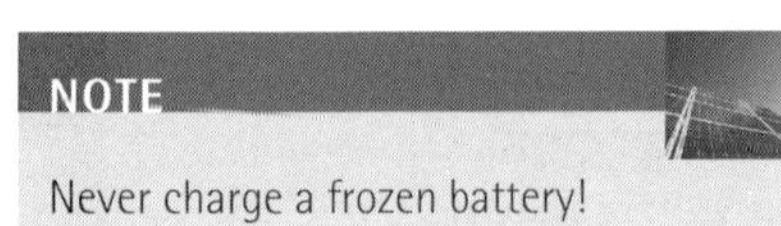

Rated Battery Capacity

This is the maximum amount of energy that a battery can produce at a constant current draw. The same discharge rate must be used when comparing battery capacity. For PV, use the 20-hour rate. This is a reasonable rate for all the variations in battery use in an off-grid situation.

Battery Life

A battery's lifetime is very difficult to predict because it depends on so many factors. These factors may include the charge and discharge rate, the number of charge and discharge cycles, and operating temperature extremes.

Sizing a Battery Bank

The purpose of the battery bank is to store energy so that power is available at night and during periods of low insolation. Size the battery bank for the highest daily use of energy and the month with the least amount of sunshine. In places with high summer cooling loads, you will size for the highest load month.

Most battery banks are sized to provide energy for two to five consecutive days. Hybrid PV systems can have smaller battery banks because they get energy from other sources. Other power sources include the following:

- Generators
- The grid
- Wind

Use the following calculations to help determine the size of batteries needed:

Daily Load AC Average / Inverter Efficiency + Daily Load DC Average / System Voltage DC = Average Ampere-Hours/Day

Average Ampere-Hours/Day × Days of Autonomy / Limit of Discharge / Capacity Battery AH = Batteries in Parallel

System Voltage DC / Battery Voltage = Batteries in Series × Batteries in Parallel = Batteries Total

Wiring a Battery Bank

The battery bank in a PV system is what stores the voltage for later use. Wiring in a series or parallel configuration was discussed previously in this chapter. It is important to note that wiring in series boosts the voltage. Wiring in parallel increases the amp-hour storage capacity of the battery bank. PV designers combine the types of wiring to create the optimal battery bank size.

Wiring a battery bank properly can be expensive. In low-voltage and high-current situations, and when current has to be conducted long distances, wiring can be very costly. Use copper wire when wiring PV systems. This is the most cost-effective and efficient way to do it.

Aluminum wire is less expensive and can sometimes be substituted for copper. However, this is not advised and is against code in residential systems. For instance, aluminum wiring can be used in long runs. Copper wiring and aluminum wiring cannot be interconnected. Special connectors are required when aluminum wiring is used. Also, aluminum has a higher resistance, which means higher wire losses for the same size wire.

You should minimize resistive losses between the batteries and the arrays to less than 5 percent for moderate-performing systems. Losses between the DC loads and boards should be kept at 5 percent or less. There should be no more than 2 percent loss between the DC control board and the battery. You will want to cut those losses by more than half for high-performance systems. The increase in conductor size will be paid for in increased performance.

Do not exceed the current rating of wires or cables. This causes them to overheat and insulation to break down. This, in turn, causes fires.

You should select wires based on system voltage, drop in voltage, and current-carrying capacity. Overcurrent protection devices should be adequately sized.

Wiring size can be noted in AWG gauge units, inches, or millimeters (mm). Stranding is sized by the number of strands and their diameter. The larger the wire, the more current it can carry.

You must protect wiring from all hazards, the environment, rodents, and people. You can accomplish this by placing the wiring in conduit or a raceway.

CHAPTER 6 SUMMARY

This chapter is a comprehensive explanation of batteries in PV systems. It describes different types of batteries available on the market. The PV designer should consider different types for different applications. Everything from battery types and classifications, to battery parts, to battery selection and installation is explained. Safety protocols for locating batteries and battery handling are also discussed.

KEY CONCEPTS AND TERMS

Ampere (A or amp)

Ampere-hour (Ah)

Battery

Battery capacity

Battery cell

Battery cycle life

Battery self-discharge

Charge

Cycle

Deep-cycle battery

Lead-acid battery

Nickel-cadmium (NiCd) battery

Sealed battery

State of charge (SOC)

CHAPTER 6 ASSESSMENT

Energy Storage Device System Design

1. Lithium batteries are the best choice for use in PV systems.
   - ❑ **A.** True
   - ❑ **B.** False

2. Which of the following are advantages to absorbed glass mat batteries? (*Select three.*)
   - ❑ **A.** Lower maintenance
   - ❑ **B.** Tolerates low temperatures
   - ❑ **C.** Can be installed in any orientation
   - ❑ **D.** Inexpensive to purchase

3. Which of the following are disadvantages of lead-acid batteries? (*Select two.*)
   - ❑ **A.** Limited availability
   - ❑ **B.** High maintenance
   - ❑ **C.** High water loss
   - ❑ **D.** Poor deep-cycle performance

4. Which of the following is the safest method when mixing an acid with water?
   - ❑ **A.** Pour the acid quickly into the water.
   - ❑ **B.** Pour the water slowly into the acid.
   - ❑ **C.** Pour the water quickly into the acid.
   - ❑ **D.** Pour the acid slowly into the water.
   - ❑ **E.** None of the above.

5. Which of the following statements are true about batteries connected in a series?

   ❑ **A.** The negative terminals are connected to each other and the positive terminals are connected to each other.

   ❑ **B.** The voltage of the batteries is the sum of all the voltages of the batteries.

   ❑ **C.** The capacity is the capacity of the lowest battery capacity rating.

   ❑ **D.** B and C only.

   ❑ **E.** None of the above.

6. Battery sizing is:

   ❑ **A.** the capability of a battery system to meet the load demand with no contribution from the photovoltaic system.

   ❑ **B.** ordering batteries to fit into the space allotted for batteries.

   ❑ **C.** sizing each module with its own battery.

   ❑ **D.** sizing each array with its own battery.

7. The purpose of the battery bank is to store energy for use during periods of low insolation.

   ❑ **A.** True

   ❑ **B.** False

8. Wiring in _______ increases the ampere-hour storage capacity of the batteries.

   ❑ **A.** string series

   ❑ **B.** aluminum

   ❑ **C.** copper

   ❑ **D.** parallel

9. Rated battery capacity is:

   ❑ **A.** the maximum amount of energy that a battery can produce at a constant current draw.

   ❑ **B.** the minimum amount of energy that a battery can produce.

   ❑ **C.** the rate at which a battery can be discharged.

   ❑ **D.** the rate at which a battery can be charged.

10. Wiring should be selected based on which of the following? (*Select three.*)

   ❑ **A.** System voltage

   ❑ **B.** Drop in voltage

   ❑ **C.** Current-carrying capacity

   ❑ **D.** Electrolyte used in the battery

   ❑ **E.** Number of modules in the PV system

Charge Control and Maximum Power Point Tracking (MPPT)

THIS CHAPTER EXAMINES CHARGE controllers and Maximum Power Point Tracking (MPPT), and why it is important to understand their use, functions, and applications to maximize battery life and PV system performance. This chapter also defines charge controllers and their functions. Next, charge-controller sizing, selection, and design are discussed in detail. Finally, the chapter explores the function of MPPT in grid-tied applications.

Topics & Concepts

This chapter covers the following topics and concepts:

- Battery charge controllers in PV systems
- Charge-controller terminology and definitions
- Charge-controller designs
- Maximum Power Point Tracking (MPPT) controller features
- PV controller system design
- Voltage regulation set-point selection
- Charge-controller selection
- Sizing charge controllers
- Operating without a charge controller

Goals

After completing this chapter, students are expected to be able to:

- Explain the flexibility, functions, and features of charge controllers
- Understand charge-controller terminology and definitions
- Discuss what it means to operate a system without a charge controller
- Recognize charge-controller designs
- Understand how to size charge controllers in grid-integrated photovoltaic systems
- Determine the requirements for charge control in battery-based photovoltaic systems, based on system voltages, current, temperature, string sizes, charge, and discharge rate
- Understand how to size charge controllers in battery-based photovoltaic systems
- Discuss the role of a charge controller within the PV system
- Understand MPPT and its role in system sizing and performance

Battery Charge Controllers in PV Systems

The charge controller in a standalone PV system regulates the charging and discharging of batteries. **Discharge** is the withdrawal of electrical energy from a battery. The charge controller controls battery charging, keeping batteries from being overcharged by the array. The charge controller also keeps the batteries from being too deeply discharged by the load.

Batteries that are correctly charged get the most out of their storage capacity. Having correctly charged batteries prolongs battery life and keeps the PV system working at its best. Charge controllers monitor temperature and voltage loads. They do this by using different algorithms. Many use meters and alarms to record and alert operators of battery-charging issues.

Charge controllers use voltage controls, current controls, and set points to prevent batteries from being overcharged during times of high insolation. Charge controllers also prevent overcharging of batteries when battery capacity is reached in the charging cycle. Damage to overcharged batteries is caused by the following:

- Overheating
- Voltage levels too high for the state of charge
- Acute gassing

- Faster grid corrosion
- Loss of electrolyte

Charge controllers prevent batteries from being too deeply discharged. Over-discharging happens when insolation is low or load usage is too high. Batteries that are repeatedly over-discharged lose capacity. Also, their life is shortened.

A series of chemical reactions occur in a deeply discharged battery. This affects the battery's ability to recharge to full capacity. In addition, overcharging prevents batteries from fully servicing system loads. Overcharging weakens the bonds between the grids and the electrolyte.

Charge controllers can switch critical loads to other electrical sources when batteries reach a certain state of discharge. This is called a **low voltage disconnect** (LVD) set point. You should include monitoring and an alarm system on the charge controller in standalone systems. This will alert the operator of low battery capacity. A system for managing and controlling battery charging increases system usability.

Charge-Controller Terminology and Definitions

Following are explanations of methods that charge controllers use. Charge controllers limit charge to and from batteries in PV systems. Other important terms

This Outback Power Flexware panel includes two 3,600-watt inverters and two MX 80 charge controllers mounted on the left side of the panel.

Courtesy of Outback Power

are also defined. Understanding these terms helps the PV designer design a PV system with battery backup.

- **Array–to-load energy ratio (A:L)**—The array–to-load energy ratio (A:L) represents the average daily usable amount of energy output by the array. This is compared with the average daily load during the worst-case month of the year. Depending on location, that can be summer or winter.
- **Voltage-regulation (VR) set point**—The voltage-regulation (VR) set point is the maximum voltage that the battery is charged based upon the specific temperature of the battery or batteries. The charge controller automatically sets and controls this feature. The charge controller stops the battery from being charged when the VR set point is reached. In some systems, the VR is set to allow the battery to overcharge slightly. This provides some gassing and battery equalization. Temperature and battery type, as well as electrolyte, affect VR. A higher VR may be required for optimal battery performance in standalone systems.
- **Array-reconnect voltage (ARV) set point**—The array-reconnect voltage (ARV) set point is the point at which, after the array has been disconnected and stops charging the battery and battery voltage drops, the battery is reconnected to the array to continue a charge cycle. You should set the ARV set point at a level that allows proper battery charging.
- **Voltage-regulation hysteresis (VRH)**—The voltage-regulation hysteresis (VRH) is the voltage difference in the form of a lag between the VR and the AVR. The VRH is critical to recharging the batteries. If it is set too high, the batteries will not have adequate time to recharge. If it is set too low, the recharging cycles take place too quickly.
- **Low-voltage disconnect (LVD) set point**—The low-voltage disconnect (LVD) set point is the voltage point where a charge controller disconnects the load from the batteries. This prevents overdischarging, which prevents the batteries from being too deeply discharged.
- **Load-reconnect voltage (LRV) set point**—The load-reconnect voltage (LRV) set point is the voltage point where the load is reconnected to the battery. The LRV set point is set at the voltage where the battery is recharged sufficiently for operation. It must be set high enough, with enough voltage difference, so that the LVD and the load will not go through rapid cycling.
- **Low-voltage disconnect hysteresis (LVDH)**—Low-voltage disconnect hysteresis (LVDH) is the difference (voltage lag) in voltage between the LVD and the voltage where the load is reconnected to the battery. The LVDH set point must be very accurate to avoid damaging the controller or the batteries by rapid cycling.

A charge controller's level of sophistication is defined in part by the preceding terms. The larger and more complex the PV system, the larger the batteries and the charge controller will be. Installing alarms on the controller helps the owner understand what to do, and when.

Charge-Controller Designs

There are five major types of charge controllers:

- Shunt controllers
- Single-stage series controllers
- Diversion controllers
- Pulse-width–modulation (PWM) controllers
- Maximum Power Point Tracking (MPPT) controllers

Shunt Controllers

A shunt controller is for use with small systems. **Shunt controllers** prevent overcharging by shunting the PV array when the battery is full. The shunt-controller circuits monitor the voltage of the battery. They switch the PV's current through a transistor when a full-charge value is reached.

Shunt controllers have an electronic part called a blocking diode. A **diode** stops a current from draining back from the batteries to the PV array at night. The blocking diode serves as a one-way valve, enabling current to flow into the batteries while they are charging, but also preventing current leakage or backflow to the array.

Shunt controllers are inexpensive and simple designs.

Single-Stage Series Controllers

A **series controller** interrupts the charging current by open-circuiting the PV array. The control element is in series with the PV array and battery. A single-stage series controller prevents batteries from overcharging. It does this by switching off the array when the battery voltage reaches a set value. This set value is the charge-termination set point (CTSP). The battery and array reconnect when the battery charge drops to the charge resumption set point (CRSP).

A single-stage series controller uses a transistor or relay to break the circuit. This breakage, or opening of the circuit, prevents the reversal of current flow at night. The controllers, which are small and cheap, eliminate the need for heat sinks because they do not produce much heat. They have a greater capacity to handle larger loads and they do not require much ventilation.

Diversion Controllers

Diversion controllers regulate the charging current depending on the state of charge of the battery. They divert extra charge current to a resistive load. The full current from the array is allowed to flow when the battery is at a low state of charge.

The controller and the load resistors reduce some of the battery power. This makes it so that less current flows into the battery when the battery gets close to full charge. The charging current drops when the battery bank reaches a fully charged state.

This type of controller works with other non-PV charging sources. The dissipation of energy creates heat, so the controllers need ventilation. The controllers usually do not have reverse-leakage–prevention systems. Therefore, the PV array may need additional diodes.

Pulse-Width–Modulation Controllers (PWM)

Pulse-width–modulation (PWM) controllers were the most popular choice for home systems. They provide a gradually tapering charge. They do this by quickly switching the full charging current on and off when the battery is close to being fully charged.

The length of the current pulse decreases slowly as the battery voltage increases. This reduces the current into the battery. PWM controllers usually include a built-in method that eliminates back-feed losses at night. This eliminates the need for blocking diodes.

Maximum Power Point Tracking (MPPT) Controller Features

Understanding what maximum power point tracking (MPPT) is and how it works is critical to effective system operations. MPPT is found in charge controllers and grid-tied inverters. A controller that has the MPPT feature enables the controller to track the maximum power point (MPP) of the array throughout the day. It then continually adjusts the MPP by electronically forcing the string or combiner voltage and current to the MPP for energy production. This enables the maximum amount of energy to be delivered to the batteries and the rest of the system.

Certain situations occur when the MPPT is not available. The array voltage changes to just above or below the battery voltage. This happens when charging the battery in high-temperature situations. Without MPPT, the array would operate around the voltage of the battery with little flexibility. This direct operating status results in a loss of array power when it cannot keep the voltage above the battery voltage for charging when the temperature on the array rises.

MPPT is beneficial, but it depends on the temperatures of the array, the string voltage, and the battery state of charge. There can be a loss of power and efficiency if the array is functioning in high temperatures with low string voltage. Many MPPT inverters enable you to size strings to not exceed 150 volts of direct current (VDC) during the maximum of the coldest conditions. This provides incredible system flexibility. This means that with some charge controllers, you can configure strings to provide higher input voltages that always keep the voltage above the charging voltage. Overall, the addition of the MPPT feature to the controller results in an additional energy gain of 10 to 30 percent. This depends on environmental conditions and proper string sizing.

FIGURE 7–1 illustrates the I-V curve. When specifying a MPPT controller, consider the following:

> **NOTE**
>
> Oversizing and undersizing strings results in system energy waste and loss. This is because DC power may be available that cannot be converted to AC power due to the MPPT conversion algorithm in PV systems. We call those limits the MPPT voltage and current window.

- **Battery voltage**—The output voltage of the controller must be above the nominal and required battery-charge voltage.
- **Maximum array voltage**—MPPT controllers have a voltage and current window that limits the allowable usable input and output voltage and current from the array. Do not exceed the maximum voltage, current, and wattage rating of the controller. See the section "Determining the Maximum Array Voltage" for help determining this number.
- **Maximum output amps**—Usually, the controller input current is less than the output current. Charge controllers have a maximum input voltage and current rating specified by the manufacturer and/or the listing agency. You

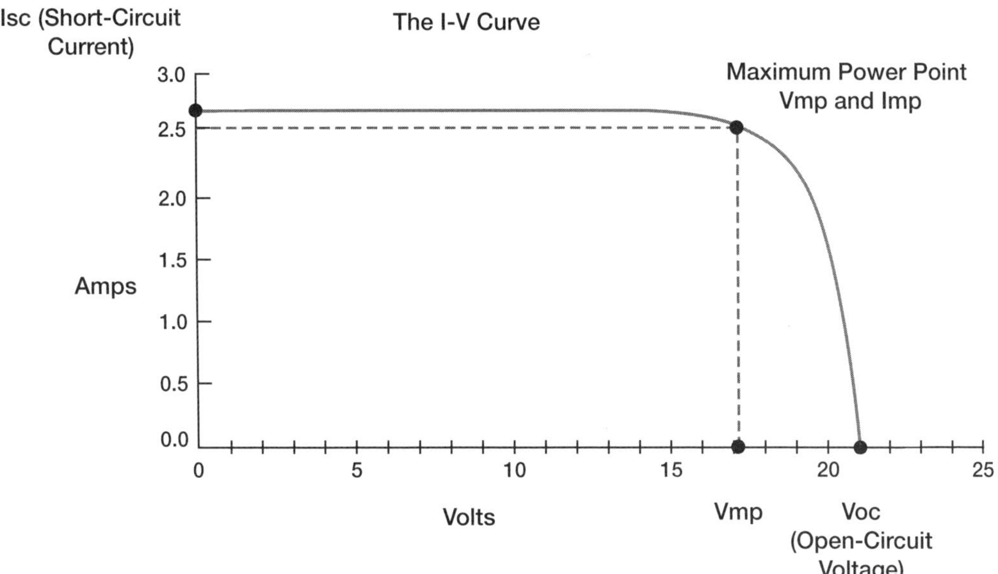

FIGURE 7–1 The MPPT window within the NEC code window. For high performance design, you must design with strings that perform within the MPPT window, which is smaller than the NEC code window.

Courtesy of D.L. King, W.E. Boyson, and J.A. Kratochvil, Sandia National Laboratories should size the controller to prevent the PV array from generating voltage or current that goes over the controller input voltage and current limits. The rated continuous current (which may be specified as input current or as output current) must be at least 125 percent of the array short-circuit output current.

- **Maximum array watts**—Manufacturers often rate their controllers by giving a specific number on the maximum array STC watts allowed at a given nominal battery voltage. This should, but does not always, match up with the MPPT conversion-window algorithm and may be limiting.

- **Optional features**—The designer should decide on a controller that has the features that will maximize the benefits of the particular system's performance. This will guide most system designs toward PWM and even more to MPPT.

Determining the Maximum Array Voltage

To determine the value, first multiply the open-circuit voltage (VOC) of the modules by the number of modules in the series. Then multiply the product by a temperature multiplier listed in Table 690.7 of the NEC 2011. (Refer to that table for determining the multiplier.) Next, derate the array properly for temperature and other conditions. You will miss a small or large part of the MPPT window if you don't. This is not a code issue; it is a design-driven performance issue.

Hitting the right multiplier means obtaining more usable energy. The Code requirements are more flexible and broad than performance design requirements. You have more string choices under code. You have fewer string choices to produce more energy when designing for performance.

The voltage and current window is delimiting beyond code purposes. Any voltage and current combination that does not fall within the MPPT performance window will not be usable to charge batteries or, in grid-tied systems, to be converted into AC power. Like a funnel, it can pass only so much through it. By understanding and applying the MPPT window in your designs, you will increase your system output.

Maximum Power Point Tracking (MPPT) and Its Impact on Systems

Good MPPT selection and proper application provides more energy to the batteries and load. It is also more cost effective over time. It boosts voltage, so that even with high temperatures on the panels it will allow high enough DC voltage to charge batteries. Another advantage of using an MPPT controller to raise voltages is that it allows the use of smaller conductor size.

PV Controller System Design

Charge controllers are used in PV systems to control the charging of batteries. Charge controllers constantly monitor voltage to prevent batteries from being overcharged or too deeply discharged.

Most charge controllers are rated at 40 or 60 amps. Amps are the amount of current a charge controller can accommodate from an array. You should size controllers to handle a minimum of 125 percent of the array's maximum output. Additional capacity can also be considered in sizing a charge controller or controllers for a standalone system for both expansion and performance purposes.

Important considerations include the following:

- The array-to-load ratio (A:L) is one of the most important factors in system design. It has a major impact on battery life, energy availability, and system performance.
- The A:L represents the average daily usable amount of energy output by the array as compared to the average daily load during the worst-case month of the year. The array component is the derated amount that can be delivered to the load or batteries. This needs to be properly adjusted for the system and environment. In some locations, that will happen during winter, and in others, during summer.
- An A:L ratio of 1.1 has long been considered sufficient for most systems. This creates a challenge for the array to recharge the batteries of the load as rated in a reasonable amount of time. A publication from the Sandia National Laboratories reported, "Test results indicate that batteries cycled in PV systems with an A:L of 1.1 yielded between one-half and one-quarter as many cycles as batteries operated at an A:L ratios of 1.3 and higher" (from FSEC/Sandia "PV Battery Life Cycle Test Procedure").

Any design element that reduces component life by 50 to 75 percent is critical. Understand this issue and avoid it in your design. When you provide five days of autonomy, three to four weeks could be required to bring the batteries back to an appropriate SOC. This creates design and user issues because the system is at risk based on design and environment.

In the design process, addressing proper battery charging is a serious issue. It can be resolved by the following:

- Increasing the ratio and cutting the battery-recharge time
- Reducing load

- Providing an additional form of electrical generation to keep batteries healthy

Increase the array size to increase recharge rate. This may require more panels or strings so that the batteries recharge in one to three days. Recharging quickly maximizes the battery capacity and life. Remember, insolation is less in cloudy winter conditions. A substantial increase in array size may be needed to achieve the right ratio.

Better energy planning and usage by reducing loads system is an option. Long low-insolation periods will probably preclude doing so for long, however.

- One approach is to add another form of energy production like a generator, wind turbine, or small hydro turbine. These may be the simplest and least expensive methods. They allow flexibility in delivering additional energy. They also shorten low-voltage battery periods. You can accomplish this by considering the following. You can maintain a higher available state of charge by sizing the charger properly. Size the battery array to get the maximum charge from the array or other source to the batteries. The peak energy collection period is from 8:00 a.m. to 4:00 p.m. solar time. This peaks at or about solar noon. The charge controller needs to match or exceed output as close to the peak input as possible, not just the average.
- There are advantages of using an MPPT controller to raise voltages and lower conductor size. You can reduce system losses from the array to the batteries by properly sizing cables and charge controllers. To do it right, you may need to add a charge controller or move to a larger one. As a component cost versus throwing away energy, this is usually the right choice in design.
- Cable runs add resistance. Your design should keep components like batteries, chargers, and inverters as close together as possible.
- A full understanding of the loads is important. Provide guidance that will level the loads. Eliminate as many large peak current draws during the operational day or week. Do this by placing some items on automated switches, timers, or alarms. Monitoring enables owners to see how quickly they are drawing current from the available energy. When owners understand these consequences, they can manage their energy use more effectively.
- Understand controller set points for the selected batteries' charging algorithms. Selecting the wrong points for voltage or current can be costly. Always protect the equipment from anyone or anything that might reset the points you selected.

■ Use of a temperature-compensation sensor is critical to maximize charge and battery life. The charge controller needs to address the battery temperature, not the internal temperature of the controller or an assumptive number. Fine-tuning will more than pay for itself. It enables the owner to see the real capacity of the battery array in real working conditions.

Voltage-Regulation Set-Point Selection

Charge controllers can range in design from complex to simple. It is very important to determine the correct voltage-regulation set point. Incorrect set points will cause batteries to overcharge, never fully charge, or discharge too deeply. This shortens battery life. In addition, the PV system will not work optimally. Correct set points will ensure batteries charge fully during insolation without overcharging.

Manufacturers typically recommend charging batteries for longer periods (floating) than the PV system allows. Float charging limits gassing and requires an uninterrupted supply of power. There are relatively few daily insolation hours available to charge the batteries.

The set point must be set to maximize voltage and current influx. This allows the array to charge the batteries in a short time period. Consider battery type when determining appropriate voltage-regulation set points.

TABLES 7–1 and **7–2** will help you determine some sample voltage-regulation set points. Different manufacturers have different charge specifications that must be followed.

TABLE 7-1	VOLTAGE-REGULATION SET POINTS FOR INTERRUPTING ON-OFF CONTROLLER AND BATTERY TYPE				
		Type of Battery			
Design type controller	Regulation charge voltage at 25 degrees Celsius	Flooded lead-antimony	Flooded lead-calcium	Sealed, valve-regulated lead-acid	Flooded pocket plate nickel-cadmium
Interrupting on–off	Per nominal 12V battery	14.6 to 14.8	14.2 to 14.4	14.2 to 14.4	14.5 to 15.0
	Per cell	2.44 to 2.47	2.37 to 2.40	2.37 to 2.40	1.45 to 1.50

TABLE 7-2	VOLTAGE-REGULATION SET POINTS FOR CONSTANT-VOLTAGE PWM, LINEAR CONTROLLER, AND BATTERY TYPE				
		Type of Battery			
Design type controller	Regulation charge voltage at 25 degrees Celsius	Flooded lead-antimony	Flooded lead-calcium	Sealed, valve-regulated lead-acid	Flooded pocket plate nickel-cadmium
Constant-voltage PWM, linear	Per nominal 12 V battery	14.4 to 14.6	14.0 to 14.2	14.0 to 14.2	14.5 to 15.0
	Per cell	2.40 to 2.44	2.33 to 2.37	2.33 to 2.37	1.45 to 1.50

Charge-Controller Selection

Charge controllers can be simple or complex. Complexity depends on the PV system's application and the designer's and owner's desire to evaluate the system status accurately. Operating a system with insufficient monitoring is risky. It is like flying blind.

Deciding which charge controller to use for a PV system depends on several factors. Consider the following when choosing a charge controller for a PV system:

- Availability, cost, and warranties
- System voltage
- Load currents and PV array
- Type and size of the batteries
- Switching element design and regulation algorithms
- Regulation and load-disconnect set points
- Surge-protection devices and disconnects
- Operating conditions
- Mechanical design and packaging
- Meters, alarms, and other system indicators

Sizing Charge Controllers

Several factors determine the charge controller's size. The most important are the array size, application, and environment. Controllers should be sized for the daily voltage and currents, and for electrical surges under the extremes of the system, not just the averages.

Arrays can produce more current than what they are rated to produce. Higher current production occurs under a variety of circumstances, especially when it is cold outside. Reflection from snow, water, and clouds can increase the nominal STC 1,000W/m² by a factor of up to 1.4.

Undersized charge controllers fail, and they make batteries fail. Replacing a failed charge controller is more expensive than installing an effectively oversized charge controller in the first place. Doing it right the first time costs a fraction of what it costs to replace the battery array.

Multiply the array's peak-rated current times the aforementioned enhancement factor (STC 1,000W/m² by a factor of up to 1.4) to determine the size of the charge controller. The array's current is the number of modules in parallel times the module current. Use short-circuit current instead of the maximum power current. If you use short-circuit current, then shunt controllers used in short-circuit circumstances should be included in the sizing calculation. This gives you a higher safety factor.

Operating Without a Charge Controller

It is safe to say that virtually every standalone PV application that is professionally installed by a licensed contractor and that is required to be reliable will need a charge controller. Are there PV systems that don't? Yes. But they are probably not the type of system on which you would wish to rely long-term for an off-grid home or cottage.

A power system built to run a reverse-osmosis (RO) water-purification system. The RO system was sent to Jordan to purify water. The system consists of six Trace charge controllers, four 5-kW Trace inverters, and 24 batteries.

Courtesy of DOE/NREL

CHAPTER 7 SUMMARY

Understanding and being able to select and program the right charge controller for the job is important to system operations, reliability, and performance. Improper selection or usage can be very expensive and quite disconcerting to the owner who wants reliable power and does not want to replace batteries in one-half or even one-quarter of the expected lifetime.

Understanding the owner's goals and objectives for his or her PV system requires some education for the owner and for you. Researching the right components and understanding the causes and effects of various choices results in better quality and a lower overall system cost. Products change and improve. Some go away, and some may be the best choice. Spend time to understand more than simple sizing. Learn why you need to make specific choices to get the results you are seeking.

KEY CONCEPTS AND TERMS

Array-reconnect voltage (ARV) set point	Low-voltage disconnect hysteresis (LVDH)
Array-to-load energy ratio (A:L)	Series controller
Diode	Shunt controller
Discharge	Single-stage series controller
Load-reconnect voltage (LRV) set point	Voltage-regulation hysteresis (VRH)
Low-voltage disconnect (LVD) set point	Voltage-regulation (VR) set point

CHAPTER 7 ASSESSMENT

Charge Control and Maximum Power Point Tracking (MPPT)

1. The charge controller in a standalone PV system regulates the charging of batteries.
   - ❏ **A.** True
   - ❏ **B.** False

2. Which of the following damages overcharged batteries?
   - ❏ **A.** Overheating
   - ❏ **B.** High voltage levels
   - ❏ **C.** Acute gassing
   - ❏ **D.** Grid corrosion
   - ❏ **E.** Loss of electrolyte
   - ❏ **F.** All of the above

3. Voltage and current set points are standard across most battery types and manufacturers.
   - ❏ **A.** True
   - ❏ **B.** False

4. Charge controllers use _______________ to prevent batteries from being overcharged.
- ❑ **A.** PV arrays
- ❑ **B.** voltage and current controls and set points
- ❑ **C.** temperature
- ❑ **D.** electrolyte

5. Which of the following is the voltage-regulation set point?
- ❑ **A.** The minimum voltage the battery can reach, set and controlled by the charge controller
- ❑ **B.** The maximum voltage the battery can reach at a specific temperature, set and controlled by the charge controller
- ❑ **C.** The set point where the array stops charging the battery
- ❑ **D.** The set point where the array starts charging the battery
- ❑ **E.** None of the above

6. Designing a battery-charging system only requires that you know how big the array is and the number of batteries you will use.
- ❑ **A.** True
- ❑ **B.** False

7. Incorrect voltage-regulation set points will cause batteries to be:
- ❑ **A.** overcharged.
- ❑ **B.** too deeply discharged.
- ❑ **C.** disconnected from the PV system.
- ❑ **D.** A and B only
- ❑ **E.** None of the above

8. Selecting the right A:L ratio will result in which of the following?
- ❑ **A.** More kilowatts per volt
- ❑ **B.** A longer battery life by 100 percent
- ❑ **C.** The ability to monitor battery capacity
- ❑ **D.** A longer battery life by as much as 50 percent

9. Operating a PV system to charge batteries without a charge controller is highly recommended.
- ❑ **A.** True
- ❑ **B.** False

10. The array-to-load ratio (A:L) is the ratio of:
- ❑ **A.** array size to total weight for mounting panels.
- ❑ **B.** battery size to shipping requirements.
- ❑ **C.** usable average daily energy available from the array to the amount the load requires, during the worst-case month of the year.
- ❑ **D.** the area the batteries need for installing in a battery container.

11. Maximum power point tracking (MPPT) does which of the following? (*Select two.*)
- ❏ **A.** Lengthens the solar day for energy collection.
- ❏ **B.** Forces the voltage and current produced by the system to provide the highest amount of power.
- ❏ **C.** Increases charging performance by 10 to 30 percent.
- ❏ **D.** Eliminates the need for charge-controller sizing.

12. When selecting a charge controller, you should base your selection on which of the following? (*Select three.*)
- ❏ **A.** First cost of the product
- ❏ **B.** Its capability to meet the goals and objective of the system design
- ❏ **C.** The information from the marketing cut sheet only
- ❏ **D.** The ability to monitor your battery bank in a meaningful way
- ❏ **E.** Your control over the batteries and your ability to manage them throughout their life

Inverters in System Design

THE FOCUS OF THIS **chapter** is grid-connected inverter selection and application in PV system design. The chapter also discusses wiring symbols and methods of operation.

The inverter is the brains of your PV system. If you use the correct inverter type and size, you can double the inverter's usable life while increasing system output by 25 percent or better. On the other hand, using the incorrect inverter type and size and failing to pay attention to the details may result in a reduction in power. It is all about the details!

Inverters come in different sizes, types, and outputs. Understanding how an inverter works within a PV system, how it accepts DC energy on one side and outputs AC energy on the other, is important. Even so, this topic is generally underaddressed or underappreciated.

This chapter does not discuss how an inverter is engineered in any great detail. Instead, it talks in terms of what the inverter sees, what it needs to see, and what it can do with the DC energy in the process of becoming AC power. Because we are taking a PV systems design approach, this chapter addresses the relationships between components. It provides enough basic design and application information to improve the output of the whole system so you will understand why it is being done and how to apply it.

Topics & Concepts

This chapter covers the following topics and concepts:

- Grid-connected inverters
- Grid-connected inverter types and construction sizes in various power classes
- Self-commutated inverters
- Characteristic curves and properties of grid-connected inverters
- Further development in grid-connected inverter technology
- PV inverter system design
- Wiring symbols and methods of operation

Goals

After completing this chapter, students are expected to be able to:

- Discuss the basic types of inverters used in photovoltaic systems, their functions, features, capabilities, and specifications
- Recognize grid-connected inverter types and construction sizes in various power classes
- Understand the function and purpose of an inverter
- Understand operational conditions and the impact on output
- List inverter features and specifications

Grid-Connected Inverters

Inverters used in grid-connected PV systems are different from most inverters used in standalone PV systems. Inverters used for grid-connected PV systems convert the direct current, DC, from the array into alternating current, AC, for the loads. Grid-connected inverters also adjust output to match and synchronize with the utility AC sine-wave frequency and voltage. The system experiences losses during the inverter's conversion from DC to AC. You can minimize losses by incorporating good design practices and by selecting the right inverter for the system.

NOTE

Inverter chargers not only convert DC from batteries, but also can charge batteries from the grid or an alternative energy source such as a generator.

Grid-Connected Inverter Types and Construction Sizes in Various Power Classes

PV inverters range from a few hundred watts to hundreds and thousands of kilowatts. The output is in single- or three-phase energy, which is based on the energy provided through the service entrance provided by the utility company. Inverters may be single inverters, which convert energy from the entire array. You can also use multiple inverters. Multiple inverters provide redundancy with multiple sources of power, which enables the system to operate even when a portion of the system is down. These types of choices are the result of discussions with the owner during the goals and objectives design phase. It requires good communication and a clear definition of what the owner wants and needs.

Self-Commutated Inverters

Self-commutated inverters use semiconductor elements in a bridge circuit turned off and on. Materials used are as follows:

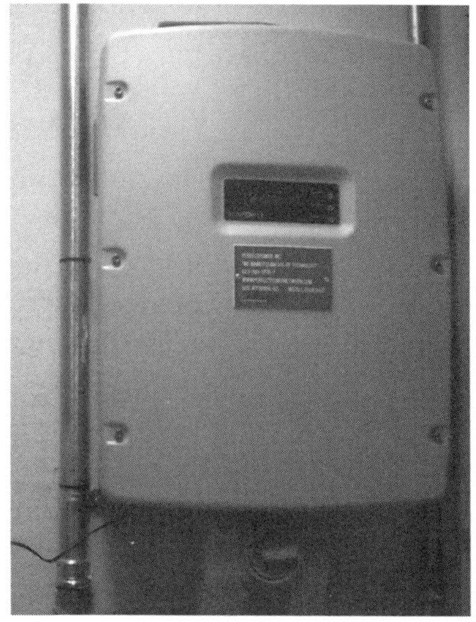

This grid-connected inverter links to the meter through the common wall. The meter is installed on the outside and the inverter is located on the inside of the building.

Courtesy of PerfectPower, Inc.

- Metal-oxide semiconductor power field-effect transistors
- Insulated gate bipolar transistors
- Gate turnoff thyristors
- Bipolar transistors

Power-switching devices use pulsation to simulate sinusoidal waves. High-frequency switching forms pulses. When used in a grid-connected system, pulsing and the resultant sine wave must match the grid's frequency. These devices are designed with protective circuits and screening devices to create electromagnetic output compatibility.

Transformers with frequencies of 10kHz to 50kHz are high frequency. They have power losses, are small in size, and weigh less. The circuitry of inverters with high-frequency transformers is complex.

Low-frequency transformers are used to match grid-controlled inverters with the grid. You can use this type of inverter for extra low voltage. The self-commutated components are as follows:

- A switching controller
- A monitoring circuit
- Maximum Power Point Tracking (MPPT)

- A grid transformer
- A full bridge

Transformer Versus Transformerless Inverters

Transformers in inverters lose energy when the inverter converts DC to AC. These losses are created by the transformer's switching devices and by its own use of energy to function. Energy used by the inverter reduces efficiency. The new inverters recover an additional 1 or 2 percentage points of efficiency due to the elimination of the transformer.

Since 2005, ungrounded PV systems have been allowed in parts of the United States after years of experience and success in Europe and Asia. The National Electrical Code (NEC) is slowly opening the door to transformerless inverters. This began in the US market under NEC 2005, and with the Magnetek Inc. Aurora PVI in 2006. The California Energy Commission (CEC), the government entity that tests and publishes the results on photovoltaic equipment, rated the inverter. This was a good step toward market acceptance.

The differences between transformer and transformerless inverters relate to efficiency, weight, functionality, and politics. Standard inverters have internal transformers that are generally heavy and operate at about 90 to 95 percent efficiency. Transformer resistances are responsible for 1 or 2 percentage points in inverter loss under standard inverter test conditions. These products have made great progress in reliability and energy production. Many have experienced solid success in the real world.

With transformerless inverter technology, there is no transformer, no need for grounding, and no additional heat generated from the transformer. There are efficiency claims of up to 98 percent, and when these inverters are fully available to the market, they will be at a lower price point—as little as half the cost of current models. This will affect system cost while boosting performance.

Challenges to transformerless inverter technology include the following:

- Transformerless inverters are not allowed in Spain, Italy, or a number of North American locations.
- There has been substantial resistance to them from utility companies.
- There is resistance from regulators and inspectors because they will require changes to local codes and standards.
- There will be a need to educate plans reviewers and inspectors on their appropriateness, safety, and reliability.
- There have been some issues with transformerless inverters related to galvanic protection.

All these challenges create obstacles to significant improvements in the industry, such as lower weights, lower costs, and higher efficiency.

Characteristic Curves and Properties of Grid-Connected Inverters

Grid-connected inverters adapt the energy as represented in the form of a current-voltage characteristic curve (IV curve) by turning solar energy into AC power. Recall that the IV curve is an instantaneous picture, graphical representation, or map of the power produced at any one point in time by the panels and strings. The units are in the following notation form: $P = IV$, where P is power in watts, I is current, and V is voltage. This energy is the DC energy as collected by the panel or module selected for the system as it is output under different

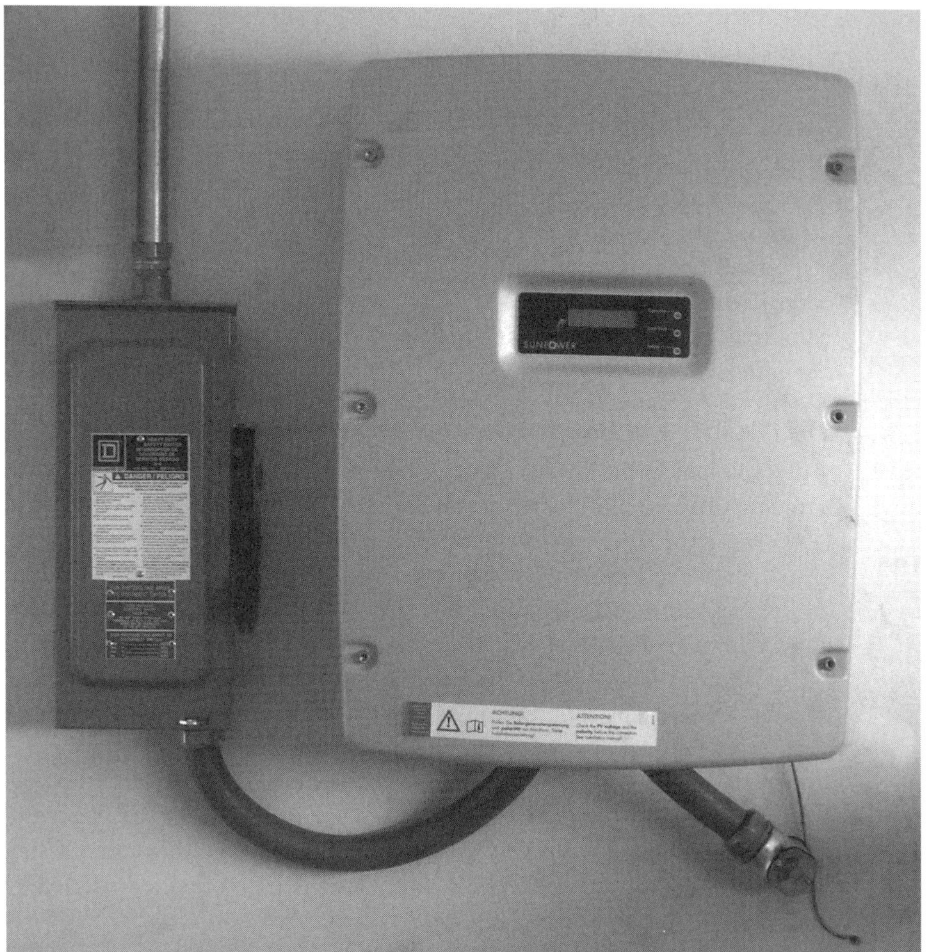

Grid-connected inverter installed at a house in Phoenix.
Courtesy of PerfectPower, Inc.

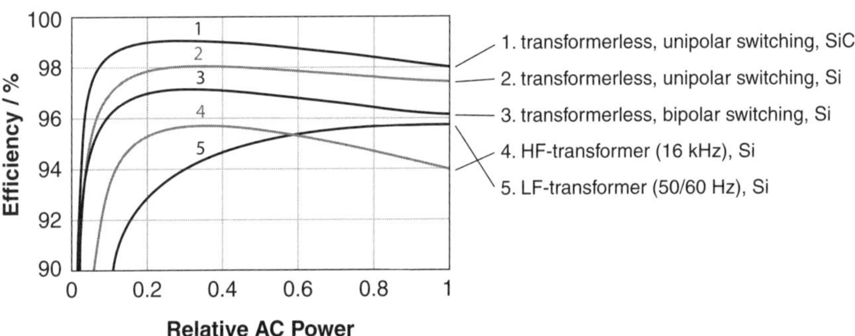

FIGURE 8-1 IV curves and performance.
Courtesy of Sandia Laboratories

conditions. Each panel type and architecture has a slightly different type of curve **FIGURE 8-1**. That means the voltage and current characteristics are different for each panel variety. These characteristics change with temperature, irradiance, and the many other factors we have discussed.

As we are dealing with grid-tied systems, we will address the role of the IV curve in the inverter design using a sine-wave inverter with maximum power point tracking (MPPT). Your goal is to get the most out of the inverter and the rest of the system.

The IV Curve and How It Affects System Performance

When you look at the IV curve for the panel presented in **FIGURE 8-2**, you will note that there is an IV curve for standard test conditions (STC) presented in black. This is the representative form that you will usually see in IV curves.

On the y-axis is current, represented as amps, which you will also see listed as I. As irradiance increases, current increases. Therefore, in the real world, the upper portion of the curve fluctuates up and down all day and all year long.

On the x-axis, you will see voltage represented as volts. On the x-axis, voltage is inversely affected by temperature. As temperature rises, the voltage is decreases; as the temperature drops, the voltage increases.

Often, the emphasis is on the IV curve at STC. People tend to burn that perfect curve into their brains. They always see it as a perfect, static curve, not as a dynamic curve. This may be why so many designers have a difficult time seeing how a PV system really works. As a result, designs are more often affected by assumption than fact.

The fact is, the STC curve is just one of millions of variations the panels will produce throughout the year. Don't panic! You will just have to get used to it and use the tools at hand to make it work for you. The fact is that you can make it

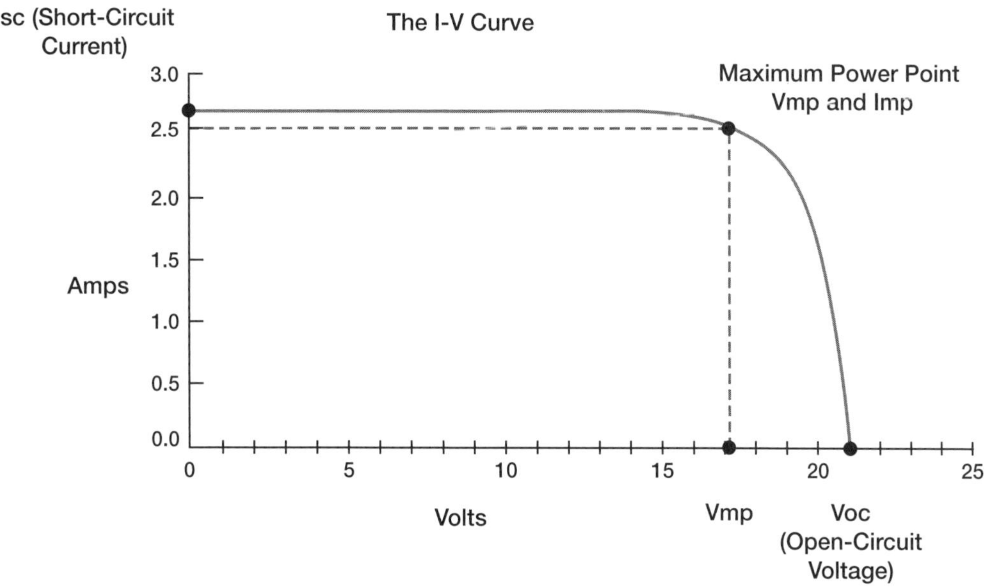

FIGURE 8-2 IV curve.
Courtesy of Sandia Laboratories

work and improve system output. You have to see it as dynamic and not a static condition of performance.

The curves generated by panels equate to a specific level of power, often seen and represented as P. Following are some of the measurements of what takes place in a PV panel as output is generated.

- **Peak power (Pmax)**—Peak power is the tested power output at STC. This is used as the panel or nameplate power rating.
- **Efficiency (η)**—Efficiency is again at STC, but it will vary when the panel is operating under any other conditions outside STC.
- **Rated voltage (Vmpp)**—Rated voltage is the voltage at STC. It is represented on the knee of the curve at the maximum power point (MPP). You will begin to see why Maximum Power Point Tracking (MPPT) is so important!
- **Rated current (Impp)**—Rated current is also at the STC maximum power point (MPP).
- **Open-circuit voltage (Voc)**—Open-circuit voltage, at STC, is where the current is a zero current.
- **Short-circuit current (Isc)**—Short-circuit current, at STC, is where the current is at the maximum voltage and the panel is literally shorted out.

- **Maximum system voltage**—Maximum system voltage is a politically determined maximum voltage allowable by code. Maximum system voltage is set in the US NEC at 600 volts DC. In Europe, however, the maximum is 1,000 volts DC. The additional voltage window allows for systems designed to that maximum and is more flexible when it comes to string sizing, while providing a slightly better inverter cost per DC watt than in the US.
- **Temperature coefficients**—These include power (P): α; voltage (Voc): β; and current (Isc): δ.

When designing PV systems, many PV designers see these numbers only as input data for string analysis. It will make a big difference in how you size strings and select inverters when you understand their full impact. They are the only real clue you will have as to how a panel, string, or system will operate under a variety of temperature conditions and ranges.

The temperature coefficients are a great clue as to how the panels will operate at higher temperatures. Now that you can begin to see graphically what the IV curve represents, the next step is to integrate it into what it means to your sine-wave inverter and how it will process the incoming DC power.

The MPPT will process energy with any number of different algorithms to force the MPP to the highest power output for the existing conditions. Remember that those conditions are continually changing with temperature and irradiance.

For example, on a sunny day, a cloud can pass overhead and change the conditions for a few seconds, minutes, or hours. This will affect irradiance, potentially boosting it by 30 percent due to lensing. Irradiance will drop 30 to 90 percent due to cloud-shade density. Temperature will also change. The MPPT scans the conditions and in the process forces the maximum power point.

You can integrate the daily and annual power generated by MPPT under all the various conditions into a series of coordinates. These coordinates on a graph indicate the MPPT performance window. You can graph that window to look like a box, which is often represented as a dotted line. We will come back to the MPPT window or box. It has an effect on what DC power will be converted into

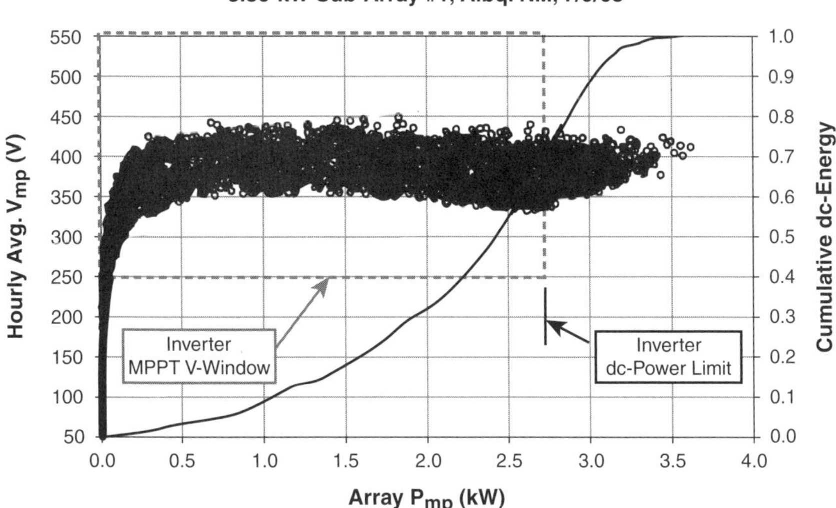

FIGURE 8-3 MPPT scatter chart.
Courtesy of Sandia Laboratory

AC, and what watt power will not. To increase performance, you want all the DC power to convert to AC without throwing any usable power away.

If conditions are changing very slowly, as they will on a cloud-free day, the MPPT does a great job continually self-correcting in small, incremental steps. This presumes a good system design, with the following:

- Balanced strings
- Consistency in tilt, orientation, etc.
- A high-quality installation

If someone has been sloppy in the design or installation, or both, the MPPT measurements will not be accurate or consistent. Therefore, the adjustments won't reflect the real situation or conditions. The inverter, the brain of your system, will not receive accurate information, and will become technologically or electrically confused. In essence, although the MPPT will attempt to do its best, it will not be able to accurately adjust for conflicts in string voltage and current because of mismatch, slope, orientation, or any other condition. The result is a system that underperforms, no matter how good the equipment is. Throw in shading, and all bets are off! The result will usually be subprime or low-moderate performance.

When daytime conditions of irradiance and temperature are changing continually and rapidly, the MPPT can't always keep up. The data it produces is sometimes behind the curve, so to speak. This data lag always throws the results off a little—and in some cases, it throws them off a lot.

When you select an inverter, add these factors to your design process, and you will see higher performance when the system operates. Faster scan frequency will provide better performance than low scan frequency. This is especially true in rapidly changing climates. Inverter manufacturers are beginning to see this as more important in their design and production. However, they are often not very forthcoming about details because of competition.

Yes, these are all complicating issues that affect output and performance. If you account for them, performance goes up, sometimes dramatically. Now that we have addressed the basics, what is the result, and what is the next step?

We have provided information about the continually changing power output. Hopefully, you will never think of the IV curve as a static process or condition.

In the Wasp diagram (See **FIGURE 8-4**), maximum power and maximum power voltage are all over the map. These normal system DC power and voltage numbers meet NEC requirements in the system tested and they meet the manufacturer's specs. They are based on a standard system design for Albuquerque, N.M. They show how performance changes—in this instance, during winter. They are not static, and they are not at STC. They have nothing to do with the inverter because they involve input power and voltage.

What if the system meets code and the manufacturer's requirements, but it is not performing well? What might the reasons for that be? One reason might be that the inverter was undersized for the system. This is one basis for our use of the

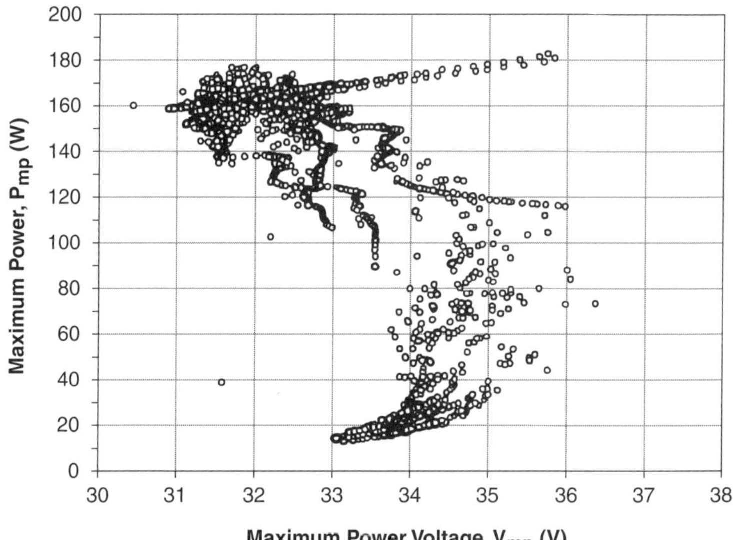

FIGURE 8-4 Wasp curve.
Courtesy of Sandia Laboratory

80 percent guideline. Based on the string sizing tool, the inverter was a good selection for the array if you are primarily designing to code.

The string sizing for the inverter allows this to take place because the string-sizing tool is tied to meeting code for safety, not performance. (Please read the introduction to the Code if you have any questions about that point.) Additionally, in time, the PV system modeling will indicate the waste condition; at this time, however, most models do not consider all the details of the MPPT window or inverter. This might be because each inverter has a separate set of window, box, or other parameters.

Other reasons might relate to the designer and the delivery chain:

There are no national or international PV design or installation standards, so the natural default is to code, which is a safety standard.

- If the designer was inexperienced, he or she may simply have gone through the motions of string sizing and design.
- If the designer was more experienced, he or she may have thought, "Yes, we will be clipping a little energy, but it is not that much. Let's not worry about it."
- The integrator owner may have thought, "We have to lower the cost of the system. It will cost less for a smaller inverter. The code and the manufacturer let me do it, so let's get the job done and get a check cashed before payroll comes due."
- The inspector, tasked with safety, cares primarily about making sure the system is safe and meets or exceeds code. He or she generally has no jurisdiction over performance. The system owner thinks he or she has gotten a great deal, and until that person realizes how the system underperforms, will remain satisfied. But there will come a time when he or she learns that something is not quite right.
- The system owner believes that PV is a commodity, which translates into equipment, design, and labor. He or she assumes they are all the same, and looks to buy the cheapest system to get the best deal.

Clipping—that is, sizing the array strings and inverter in a way that results in losing a little bit of energy for a small part of the day or year—may seem insignificant, but it isn't. It adds up over time. Don't design or install a system that clips. It will catch up to your customer—and you. If you have ethics and a conscience, it will give you a lot to ponder at night!

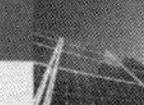

If you used to think PV is a commodity, now you know better. There's no excuse for designing low-performance or subprime PV systems. It is not that difficult to do much better as a designer or an integrator.

Further Development in Grid-Connected Inverter Technology

Described next are several grid-connected inverter concepts of which the PV designer and integrator should be aware.

Micro Inverters and Converters

String inverters are often used in PV systems that are shaded and/or in systems with modules at differing orientations. PV systems with these restrictions do experience energy losses, which can be eliminated by better design. Many companies are looking at the use of micro inverters and converters, which attach to the back of the panels. This eliminates DC strings and replaces them with AC strings. In time, they may be a functional solution for shading, multiple orientations, and other items of concern in the industry. Currently, however, many of their claims seem too good to be true, and they have not shown a consistent track record. Manufacturers do not seem to be able to back certain claims or provide data that supports their marketing tools.

Do these products work? Yes, especially in lower-temperature areas. They may allow for strings of one (when not abused), and can address shading issues.

However, if the inverter is the weakest link in the system, you have now multiplied the weakest link. As inverters fail or underperform, they will need to be replaced. There are no long-term performance records. The question of their reliability will be answered over the next few years.

Mirroring some of the challenges with existing systems is the monitoring issue. With some micro inverters, online monitoring is a separate service and is relatively expensive. Because many customers and integrators do not understand the importance of monitoring, they opt out of paying for it. As a result, simple problems are ignored until they become large and critical.

All of this will even out over time. We look forward to clear and concise information that will move these products into the mainstream of the market at the same or better cost and performance of other types of technology.

Master-Slave Inverter System

A concept called the master-slave is used for inverters beginning in the 2.0 kW range and greater, where inverters are linked together to make the equivalent of a much larger inverter. This is similar to the way strings of panels and batteries make the system bigger.

The master-slave relationship is a structure where inverters are controlled by a single master inverter. In the Outback Flex series, a single inverter is programmed to control one or more inverters. In the Fronius products, the approach is similar: Inverters are stacked in a way that allows inverter usage to be measured. They automatically shift the responsibilities of master control based on usage. This provides balanced usage over time across the inverter stack, which will give a much longer life to the products.

The advantages include the following:

- Inverters begin working as required based on irradiance, so that in the morning, one inverter will start to bring on the stack as needed.

- This setup limits inverter efficiency losses by operating at the inverter's sweet spot, which will be somewhere between 40 and 80 percent of total inverter power.
- The inverters are not underloaded or overloaded when the system is designed well.
- Thanks to redundancy, if one in four or one in 10 inverters is down, the loss is only a maximum of 25 percent, or 10 percent of the potential output, compared to 100 percent. If the master relationship is always rotating, an inverter could be down and replaced before any power is lost based on time of year or other insolation issues.
- If the inverter has strings of different orientations or outputs, it allows the MPPT to maximize them for added performance.

There is usually a three-phase option. A three-phase feed can be used to optimize power in the low-input ranges. This optimizes efficiency, keeps circuitry simple, and allows for a longer inverter life.

PV Inverter System Design

The modules in photovoltaic system collect photons and convert them to direct current (DC) that, a standalone or grid-tied inverter converts to alternating current (AC). AC is the primary type of current used in the United States and worldwide. The preference for output energy will be AC power with a clean sine wave. This means the inverter must produce a sine wave at a frequency of 60Hz with a well-shaped sinusoidal wave.

PV systems use inverters to convert DC from the arrays and/or batteries to AC. Any modern inverter on the market today can do this.

The primary questions to answer when selecting an inverter are as follows:

- How well does it make this conversion?
- How is it accomplished?
- How do you get the most out of the process?

There are many different types of inverters. This chapter concentrates on waveform inverters. Later, we will talk about transformer and transformerless inverters. There are three categories of waveform inverters: square wave, modified square wave, and sine wave. We will concentrate on sine-wave inverters for grid-tied applications.

Different types of inverters share the following standard features:

- **High efficiency**—Modern inverters convert 90 percent or more of the DC to AC under the right environmental conditions. PV designers should select an inverter that is rated as high efficiency over a broad range of loads and conditions.

- **Low standby losses**—The inverter should be highly efficient even when no loads are being generated, requiring very little standby power.
- **Frequency regulation**—The inverter should maintain 60Hz output for the North American market. Other areas around the world usually require a 50Hz output.
- **Ease of service**—The inverter should be easily maintained. When possible, select inverters with internal modular components that can be replaced easily. Look for inverters that allow sufficient space to bring conductors into the inverter for proper termination and testing. Make sure there is room to work in the cabinet, whether it is a small (less than 1,000-watt) inverter or a megawatt inverter.
- **Reliability**—The inverter should be robust and dependable, with a record of low maintenance.
- **Light weight**—The inverter should be easy to install and service.
- **Temperature characteristics**—The inverter must be able to operate well and efficiently over a large temperature range, especially in high temperatures with minimal derating. It must efficiently be able to cool its electronics for performance and extended life.
- **MPPT**—Maximum power point tracking is available in almost every grid-tied and standalone inverter product on the market. Although not all MPPT is alike, it is important to select the right product for the system.

Inverters have several optional features that clients may want to add:

- **Remote control/data monitoring**—Remote control/data monitoring allows you to control or monitor the inverter from a distance. Although this is considered an option on inverters, it is absolutely critical to proper design and service of the system and should be a standard in all PV systems.
- **Load-transfer switch**—PV systems with multiple inverters can use a load-transfer switch to transfer load to a functioning inverter in the event an inverter fails.
- **Parallel-operation capability**—In PV systems with multiple inverters, you can wire the inverters in parallel to enable them to service more load simultaneously. This is referred to as parallel-operation capability.

The Application of Inverters in System Design or How to Design Better Systems

The section "Characteristic Curves and Properties of Grid-Connected Inverters" gave you an accurate picture of what was going on with the DC-input side of the inverter. It also talked about how not to look at the IV curve as a simple, clean, static curve. It covered the issues of array size and why you should not oversize or undersize arrays. It also discussed what MPPT does, the limits of MPPT, and the consequences of not selecting the right inverter for the array. Now let's walk though a simplified process of designing a system with the right inverter for the job. We will do our best to consider most of the incremental losses that are usually ignored.

So, how big is the system going to be, and what are the local conditions?

- **System size**—4,000 to 4,500 watts
- **Location**—Phoenix, Ariz. (This is a good example of a high-temperature location.)
- **Panel type**—SunPower SPR-230-WHT
- **Orientation**—True south
- **Tilt/slope**—20 degrees
- **Maximum air temperature**—122 degrees F
- **Minimum air temperature**—14 degrees F
- **Maximum cell temperature**—~200 degrees F
- **Shade**—None
- **Solar availability**—8:00 a.m. to 4:00 p.m. (minimum)

If you prefer, you can select a local example and start with the basic information as outlined here. Either way, you will use a variety of string-sizing tools, where you will enter all the panel and inverter information and provide a great variety of output information. It will be up to you, the designer, to input the right environmental information to get the best results.

Here are a few things to consider:

- Designing for performance initially requires a lot more thought, research, and work. It requires that you consider the effects of decisions in far more detail than when designing for code. Mastering the technology and tools of your trade will set you apart from your competition in the quality of work you provide and in the respect that your peers will eventually have for you.
- As with riding a bike, as you master the design techniques, designing becomes second nature. You begin to make selections and ask questions that will give you good design solutions. It becomes easier with time—especially if you are inquisitive and follow up on the actual production of your systems.
- Most designers do not have any idea why they need to do what the Code requires. This limits their ability to move to a higher quality of design or performance. You will begin to understand the who, what, when, where, and why of better design. This is invaluable!
- Performance design will cost more, yet will produce more with lower operation and maintenance costs.

- Review a number of panels, inverters, and options to understand the differences in string-sizing applications. You will notice the information is dramatically different across the board. Now find out why, and what it means. Initially, there will be more questions than answers available.
- Ask manufacturers questions. Understand that the answers they provide will not be in the marketing literature. The industry is moving in the high-performance direction. It's better to be on the wave of change and to learn from the inverter designers who try to figure it out.
- Start with simple performance design analysis and observe the trends you see in the design options that appear.

For more information, check out the following websites:

- **Blue Oak Energy**—www.pvselect.com
- **Fronius**—www.fronius.com
- **Kaco**—www.kaco-newenergy.com
- **Outback Power**—www.outbackpower.com
- **PVPowered**—www.pvpowered.com
- **SMA**—www.sma-america.com
- **Solectria**—www.solren.com

While looking at these websites, read the manufacturers' disclaimers. Consider what you are attempting to achieve with string sizing for your specific designs.

TABLE 8-1 provides a step-by-step comparison of designing to the NEC versus designing for performance.

Other functions of the inverter are to record and store operating data and signaling system operations, and to function as system-overload protection.

Wiring Symbols and Method of Operation

Grid-connected inverters are tied either directly to the main grid or to the building's grid connection through the service entrance switchboard (SES). When the inverter is tied to the building's SES, electricity from the PV system is used by the load first. Any remaining current feeds into the main grid, where, with net metering, it can be recovered for later use.

Small residential and commercial PV systems (usually under 25 kWp) are built as single-phase systems. Larger commercial and in some locations residential systems are designed and built as three-phase systems. Those systems will primarily look the same while using different electrical calculations for wire-protection sizing and overcurrent protection.

TABLE 8-1	DESIGNING TO CODE VERSUS DESIGNING FOR PERFORMANCE	
Task or step	**Code minimum design**	**Performance design**
Complete site analysis.	May be lacking detail or may be a review of aerial images.	Detailed information collected and applied.
Determine project goals and objectives.	May be considered. Often, it is just a sizing operation.	A full discussion with the system owners or team results in goals and objectives over a minimum 20 years of system life, and preferably over a 40-year life.
Gather site environmental data.	Use local airport or other weather location data.	Develop accurate site data, including any microclimate information that will affect the system.
Determine system size.	Generally by DC watts and cost.	Determined by system output over time.
Determine how to obtain a specific required output.	Use performance model standard settings.	Use model with settings that meet all site, install, and equipment details and peculiarities.
Determine how size will produce design output and performance.	Consider derating factors as generally seen in the industry. They may not be accurate for the site or the system unless the designer understands the specifics of integrating the components, environmental responsiveness, quality of installation, and ongoing maintenance.	Break down each segment of the system and derate it. This requires establishing specific numbers while looking for areas of loss and designing around them. It also requires providing specifications and procedures to test against during and after installation to assure that losses have been minimized. It requires substantially more experience, knowledge, and information to raise the performance, yet design is where the greatest leaps in system output will be determined.
Determine panel-reducing options to one or two products.	Lowest cost or what is available.	The panel should meet the environmental, voltage, and reliability conditions. Look closely at temperature coefficients and cell architecture.
Match panel output characteristics to inverter characteristics.	N/A. Just make sure you are within code.	Review of output and temperature coefficients will either allow for the use of the panel or eliminate it.

(continued)

TABLE 8-1	DESIGNING TO CODE VERSUS DESIGNING FOR PERFORMANCE (Cont.)	
Task or step	**Code minimum design**	**Performance design**
Review string sizers.	Usually for only one product or product group. This is a moot point with many prepackaged systems.	Different inverter products will result in different string-sizing options. Some sizers offer a broad range of string options, some provide fewer. When reviewing a variety of panels on a number of different sizers, you will notice the voltage trends for the inverters, which will help you select both product and string size.
Determine sting size.	Use string sizing tool.	Modify string-sizing tool input for real conditions and eliminate low-performing options outside the MPPT window. This means you will most likely reject the high and low string sizes provided by the string sizing tool. This will vary depending upon the manufacturers' products. It limits choices of string sizes while keeping the sizing within the MPPT box. It also requires looking at options beyond the first choice and may require looking at other products, which will require more study.
	Select string size based on the greatest number of panels on the inverter that the code and the manufacturer's string sizing tool will allow.	Review string sizing tool output for inverter selection based on the 80 percent rule. An approximately 4,000 to 4,500 DC watt array will require an inverter rated at 5,000 to 5,600 AC watts. Look carefully at high and low voltage compared to temperature to make sure you are within the MPPT window for all conditions.
	Contact the manufacturer's marketing department for additional information on the MPPT window.	Establish relationship with a number of manufacturers' inverter design engineers to understand the active dynamics of their inverters, including high and low temperature efficiency, DC input maximum and minimum, and MPPT window. Request performance data for high-temperature operation.
Select potential inverter choices.	Least cost.	Inverters that meet performance and environmental requirements for your system and design.

TABLE 8-1	DESIGNING TO CODE VERSUS DESIGNING FOR PERFORMANCE (Cont.)	
Task or step	**Code minimum design**	**Performance design**
Narrow choices.	Cost and availability.	There is always a balance between cost and product selection. If customers are not educated on the facts, design choice cause and effect, or consequences, they will tend to select the least or lower-cost options. Most owners will understand the facts if they are presented with clear concise information early in the process.
Draw array layout. Use simple template or draw from scratch.		Use simple template or draw from scratch, then review the details of the system such as the combiner and inverter location for exposure to sun, ventilation, positioning of components, and wire-run options.
Draw the one-line drawing.	If it meets code, move on to draw the three-line diagram.	Review each component for code first and then address each line and component for production improvement. Have a review process where a second set of eyes reviews for errors or improvements.
Draw the three-line diagram and the rest of the set.		Have your team review the drawings to find errors and improvements. It is less expensive to get it right at this stage than to have to resubmit to the authority having jurisdiction (AHJ) or utility due to errors. It is also better to catch conductor or overcurrent-protection issues, or other potential challenges at this point than to do so during installation and have to redraw.
Draw as-built drawings.	May or may not update.	Always provide accurate as-builts to the customer, the file, the AHJ, and utility.

CHAPTER 8 SUMMARY

Inverters convert DC from the array to AC for use by the load. This is a simple concept, but for high-performance, high-output systems, it is a complex process. When selecting an inverter choose to oversize inverters in the PV systems for extended life and MPPT window compatibility. Undersized inverters work harder and create heat in the inverter and will lead to greater levels of clipping, which wastes energy.

Take your time to learn how the inverter works within the whole system and you will excel in PV system design.

KEY CONCEPTS AND TERMS

Clipping

Load-transfer switch

Maximum system voltage

Parallel-operation capability

Peak power

Rated current

Rated voltage

Remote control/data monitoring

Short-circuit current

CHAPTER 8 ASSESSMENT

Inverter System Design

1. Batteries store energy in the form of alternating current.
   - ❏ **A.** True
   - ❏ **B.** False

2. Inverters change:
   - ❏ **A.** alternating current to direct current.
   - ❏ **B.** energy from the array to the load.
   - ❏ **C.** direct current to alternating current.
   - ❏ **D.** energy from the battery to the load.

3. Which of the following are true of inverters? (*Select two.*)
   - ❏ **A.** They keep the AC from reverting to DC.
   - ❏ **B.** They record and store operating data.
   - ❏ **C.** They provide overload protection.
   - ❏ **D.** They store DC current until needed as AC current.

4. Inverters can be which of the following types of PV systems? (*Select three.*)
   - ❏ **A.** Loud to operate
   - ❏ **B.** Grid tied
   - ❏ **C.** Standalone
   - ❏ **D.** Gas producing
   - ❏ **E.** Grid tied with battery backup

5. Which of the following are standard features of an inverter?
- ❏ **A.** High efficiency
- ❏ **B.** Light weight
- ❏ **C.** Reliability
- ❏ **D.** Power correction
- ❏ **E.** All of the above

6. Master-slave setup is used for inverters in which of the following situations?
- ❏ **A.** When one inverter is the primary inverter in controlling a stack of inverters
- ❏ **B.** When insolation is very high
- ❏ **C.** When the MPP is set at high temperatures

7. The inverter needs to be efficient even though no loads are using energy.
- ❏ **A.** True
- ❏ **B.** False

8. Inverters have what types of waveforms? (*Select three.*)
- ❏ **A.** Square wave
- ❏ **B.** Sound wave
- ❏ **C.** Blue wave
- ❏ **D.** Sine wave
- ❏ **E.** Modified square wave

9. Inverter chargers are available that can do which of the following? (*Select two.*)
- ❏ **A.** Charge the array from the batteries.
- ❏ **B.** Convert DC from batteries to AC.
- ❏ **C.** Charge batteries from the grid or other AC power source.
- ❏ **D.** Use batteries to change AC to DC.

10. Inverters record which of the following operating data? (*Select three.*)
- ❏ **A.** Time of sunrise and sunset
- ❏ **B.** Input and output
- ❏ **C.** Inverter status
- ❏ **D.** Minutes of cloud cover
- ❏ **E.** Generated energy volume

Cable Wiring and Connection Systems

THIS CHAPTER EXAMINES MODULE and string labels, connection systems, cables, circuit breakers, and fuses.

Topics & Concepts

This chapter covers the following topics and concepts:

- Cable wiring and connection systems
- Conductor protection
- Module and string labels
- Connection systems
- DC cable
- AC connection cable
- DC disconnect switches and DC load switches (DC main switches)
- AC disconnect switches
- Miniature circuit breakers
- Fuses and fuse holders
- Ground fault protection

Goals

After completing this chapter, students are expected to be able to:

- Discuss the functions of module and string labels, connection systems, DC main cable, AC connection cable, direct current load switches, and circuit breakers.
- Demonstrate an understanding of the 2011 National Electrical Code articles that are relevant to solar photovoltaic systems.
- Read National Electrical Code articles concerning cable wiring or connection systems.
- Select the correct labeling and code section for various system functions.
- Discuss the importance of labeling and the consequences of mislabeling.

Cable Wiring and Connection Systems

FIGURES 9–1 and **9–2** illustrate how a system can be assembled and connected.

There are specific types of conductors and cable you should install in PV systems. Different cables are required for each section of the PV system, dealing with the following:

- Current
- Voltage
- Temperature
- Moisture and humidity
- Fully wet locations
- Conduit underground

Module and string cable differ from AC connection cables and DC main cables. Module/string cables are used outside. They connect each module to the panel junction box.

Lay positive and negative pole cables in separate compartments or conduits. Following this practice will ensure fault protection and prevent the system from short-circuiting. Use double-insulated, single-wire cable if necessary.

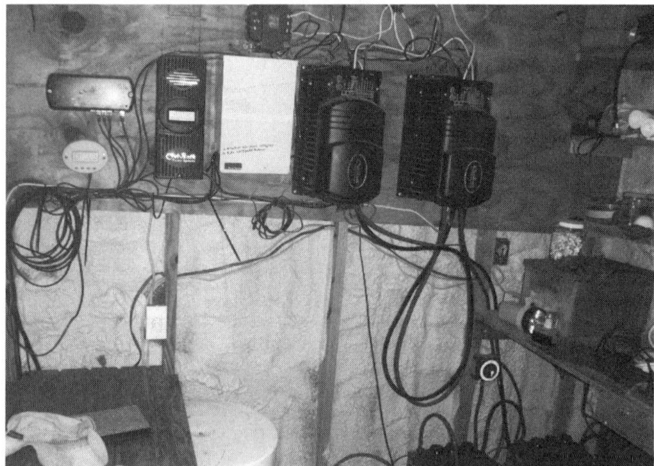

Not to code! Sloppy wiring is not only unpleasant to look at, it is very dangerous.

Courtesy of Outback Powe

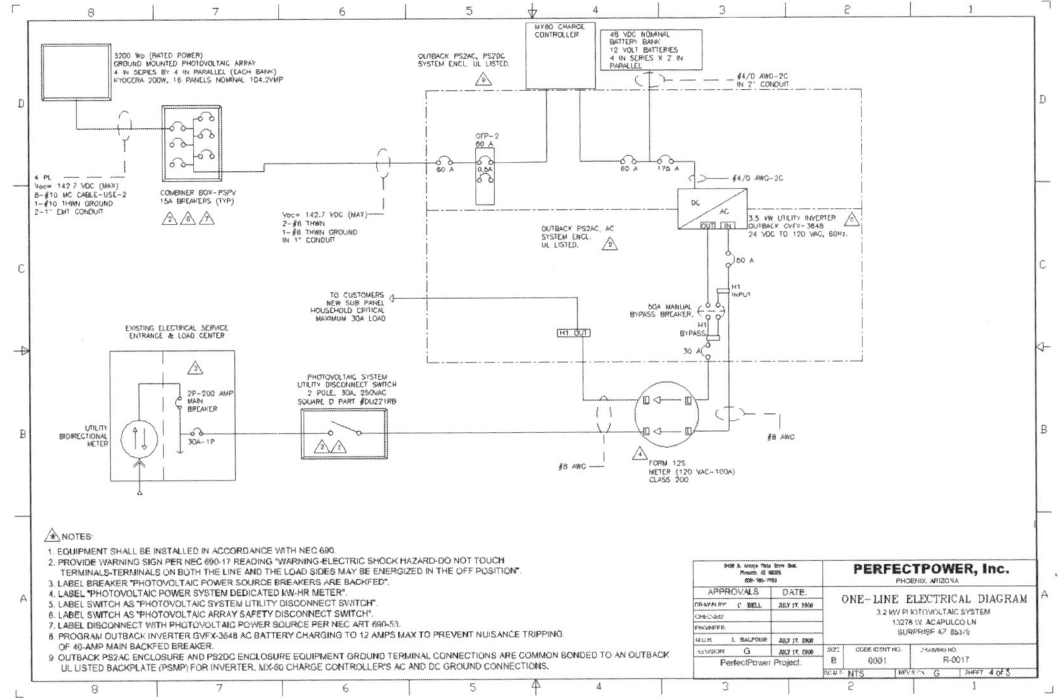

FIGURE 9–1 One-line diagram.
Courtesy of PerfectPower, Inc.

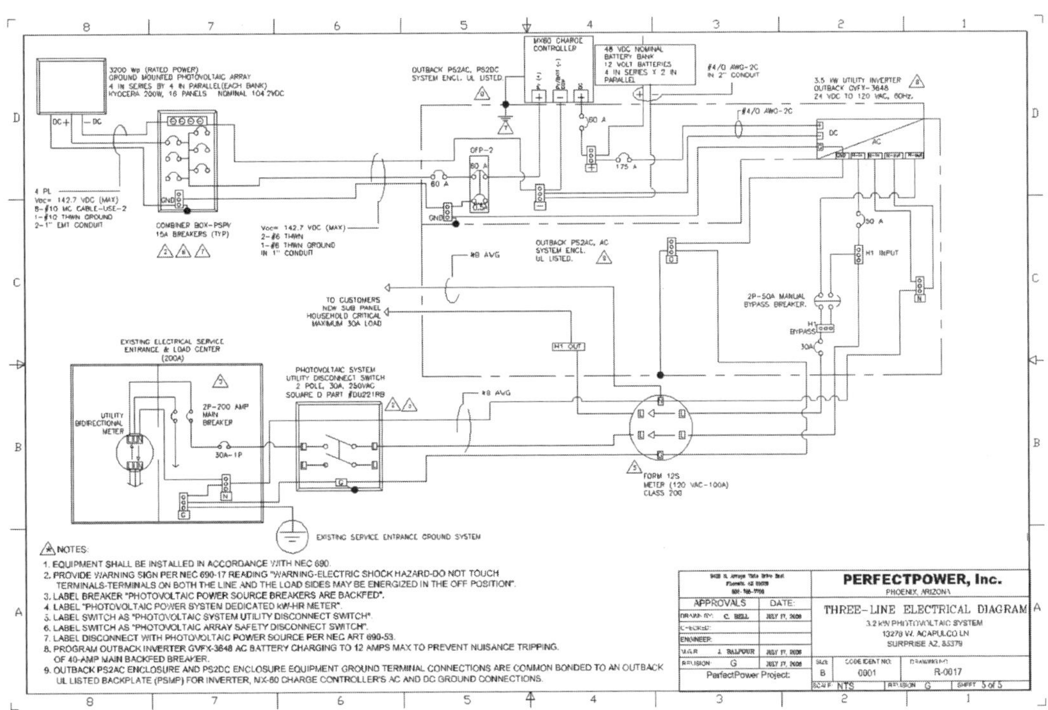

FIGURE 9–2 Three-line diagram.
Courtesy of PerfectPower, Inc.

Cable of the type H07 RN-F—a rubber-coated, double-insulated cable—can be used only in PV systems with a high operating temperature of 60 degrees Celsius. They are not suitable for tile roof installations. Tile roofs can have temperatures as high as 70 degrees Celsius.

Use cables designed for PV applications. They are designed for outside use and have the following properties:

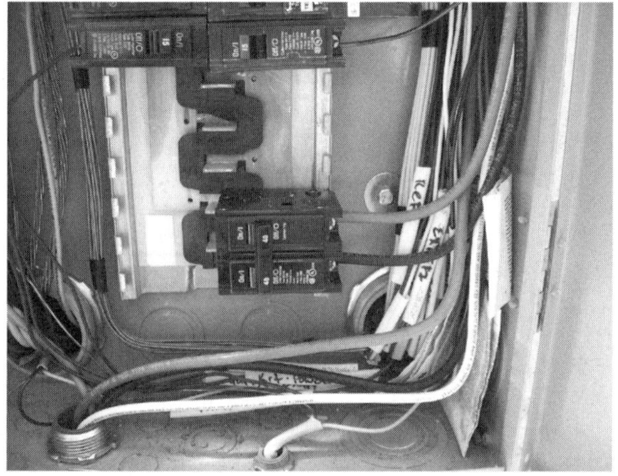

The ground conductor was bent going into the termination. Is the conductor too short, was the installer in a hurry, or did he just not care?
Courtesy of PerfectPower, Inc.

- They are UV resistant.
- They are weather resilient.
- They can tolerate a wide temperature range, from –55 degrees Celsius to 125 degrees Celsius.
- They can be encased in wire mesh, providing protection from over-voltage and from rodents.

Following are cable requirements:

- **Mechanical resistance**—Compression, bending, shear loads, tension.
- **Earth fault-proofing/short-circuit-proof installation**—Individual cable with double insulation.
- **Weather resistance**—Temperature resistance for roofs of 70 degrees Celsius or 55 degrees Celsius for lofts. Cables laid outside must have UV resistance.

Conductor Protection

Conductors are the most important component of your PV system. You should design them with safety and protection of the conductor in mind. Size conductors to meet or exceed the NEC.

The following are some pointers for providing conductor protection:

- Conductors must have wet-rated insulation.
- You should place conductors in electrical conduit.
- You should minimize wire length wherever possible.
- You should reduce abrasion and strain on the conductor by preventing it from touching the roof or being blown by wind.
- You should keep conductors in the coolest locations to reduce resistance losses.

What looks like a clean installation leaves combiner and DC conductors in the sun while leaving the cable exposed to animal damage. Even USE-2 conductors will eventually break down in the sun, risking short-circuiting and additional expense.

Courtesy of PerfectPower, Inc.

- You should oversize the conduit by one size to make wire pulling easier, as well as to reduce conductor temperatures.
- All insulated conductors must be out of the sunlight, and should also be protected from other environmental conditions.

Note the permanency of conduit installation, with waterproof attachment to the roof surface and through the parapet wall. This installation allows for easy access to the roof surface for roof maintenance for the next four decades or more while protecting the conductors from the sun and other hazards.

Courtesy of PerfectPower, Inc.

To ensure wire protection, design all conductors and conduit for expansion and contraction. This includes the following:

- All bends
- Joints
- Interfacing connectors (these may separate or loosen during expansion and contraction)
- Stressing
- Stretching

Compressing the wire can lead to insulation failure and possibly a short circuit. You must design all conductors and conduit to accommodate changes in the environment. The environment can compromise protection; this includes environmental and human-produced conditions that may cause damage. Provide additional labeling for both visible and not-so-visible identification (for example, with conductors hidden behind walls or other construction materials).

Conductors' survivability is the combined responsibility of the designer and the installer. Survivability is for the life of the system, not just the local requirement of two to five years. Survivability and safety must include the use of wire, conduit clips, and wire ties that do not break down in local environmental conditions. Overcome time and elements in the design phases. Clips must last for at least 20 years. You should refer to the current edition of the NEC for the most up-to-date requirements for conductors in PV systems.

CONDUCTOR COLOR CODES

Another aspect of conductor safety is color. In a system where one conductor is grounded, you must use conductors that:

- Are white
- Are gray
- Have three continuous white stripes of any color insulation except green
- Or are visibly marked all the way around the wire with white or gray at each termination.

Marking, versus insulation color, is allowed only on conductors size 4 American wire gauge (AWG),or larger.

If you are using conductors for module frame or metal equipment grounding, you must use outdoor-rated bare wire, or have green insulation or identification with yellow stripes. Consult NEC 2011, article 200.6, 200.7, and 322.120 for more information.

Module and String Labels

TABLE 9-1 lists manufacturers' module cable types and characteristics.

Connection Systems

Use caution when making your connections. Arcing and fire can occur because of poorly made connections. Following are four types of connections to use in PV systems:

Conductors terminated at PV system breaker in service entrance with clean smooth bend to eliminate stress on wire. It's important to install connection systems correctly. Improper installation can lead to major issues, including arcing and fire. *Courtesy of PerfectPower, Inc.*

- **Plug connectors**—You can use touch-proof plug connectors with module-connecting leads. This simplifies module installation. They're used in grid-connected PV systems. Their use is becoming common. Voltage drop is less than 0.025V.
- **Post terminals**—Cable lugs are clamped between the nut and the post to make the connection to the post terminals.

TABLE 9-1	MANUFACTURERS' MODULE CABLE TYPES AND CHARACTERISTICS		
Module cable for outside use	Nominal voltage V/$\boxed{A}$ - $\boxed{A}$	Range of temperatures	Cable cross sections (mm^2)
Flex-Sol	600V/1,000V	–40 degrees Celsius to 90 degrees Celsius	2.5, 4, 6
Laptherm Solar Plus	900V/1,500V	–50 degrees Celsius to 120 degrees Celsius	2.5, 4, 6, 10
Radox 125	600V/1,000V	–25 degrees Celsius to 125 degrees Celsius	2.5, 4, 6
Siemens Solar Cable	1800V/3,000V	–40 degrees Celsius to 120 degrees Celsius	2.5, 4, 6
Solar cable*C	1800V/3,000V	–25 degrees Celsius to 90 degrees Celsius	2.5, 4, 6
Solarflex 101	600V/1,500V	–30 degrees Celsius to 125 degrees Celsius	2.5, 4, 6, 10, 16
TECSUN S1ZZ-F solar cable	900V/1,800V	–40 degrees Celsius to 120 degrees Celsius	2.5, 4, 6, 10
Titanex 11 H07RN-F	450V/750V	–35 degrees Celsius to 85 degrees Celsius	2.5, 4

This PV combiner shows a clean installation, where many string cables are carefully protected, brought into the combiner and terminated, and then brought through to the inverter side as a reduced number of conductors. Follow NEC and local guidelines to determine the correct cable type.

Courtesy of PerfectPower, Inc.

The AC disconnect is required by code and allows the PV system to be isolated from the AC side of the building and grid. This is extremely valuable for maintenance and testing.

Courtesy of PerfectPower, Inc.

- **Screw terminals**—Metal end sleeves at the end of stranded wires are used to connect screw terminals.
- **Spring clamp terminals**—Cables can be attached without metal end sleeves when a connection box is used that is equipped with spring clamp terminals.

DC Cable

Cables listed in **Table 9-1** can be used for the DC cable. The inverter connects to the PV junction/combiner box via the DC cable. Follow NEC and local regulations in choosing the correct cable type.

Cable can be used inside PVC marked NYM or NYY. PVC is not UV resistant. If you use PV outside, put it inside conduit to protect it from UV. Try to use halogen-free PVC products to be environmentally conscious.

Use single-wire coated cables to avoid grounding faults and short circuits when installing individual positive and negative cables. In multi-wire cables, the grounding wire must not carry any voltage. Install cables where damage from rodents cannot occur.

Provide lightning and surge protection by using screen cable. All poles with the DC main cable should be able to be switched to zero potential at either the junction box or the disconnect switch.

AC Connection Cable

The AC connection cable connects the inverter to the grid through the existing AC equipment. This will include protection equipment. Three-phase inverters use a five-pole cable and connect to the grid.

Follow NEC and local regulations in choosing cables to connect the AC to the PV system. Cable types NYY, NYM, and NYCWY are allowed.

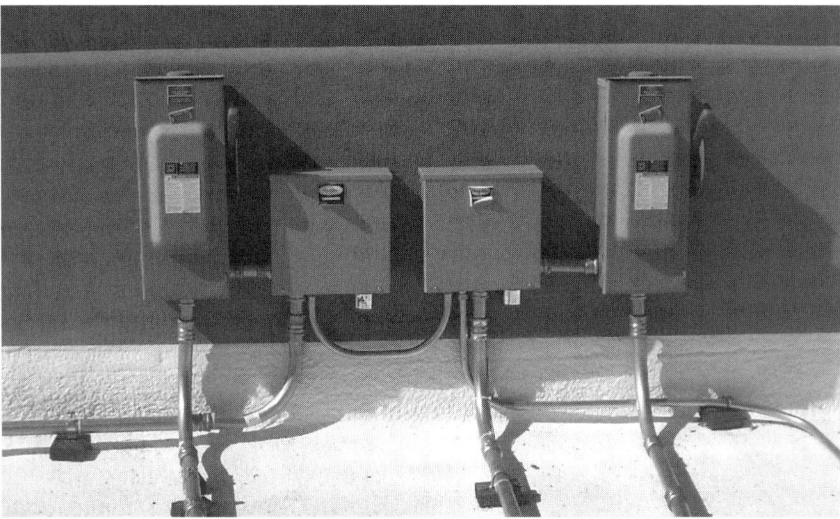

The DC disconnect switch allows both the customer and workers to isolate the inverter from the main switch.

Courtesy of PerfectPower, Inc.

DC Disconnect Switches and Direct Current Load Switches (DC Switch)

When maintenance needs to be performed, or if there is a short, it is important to be able to isolate the inverter from the PV system. The IEC 60364-7-712 standard requires that the inverter be isolated from the system.

The DC disconnect switch needs to be rated for the PV system's maximum open-circuit voltage and short-circuit current. Install the DC disconnect switch before the inverter and not in the combiner box. With some inverters, the DC switch is inside the inverter. This technology is evolving.

AC Disconnect Switch

The AC disconnect switch must have the following attributes:

- It must have a double pole.
- It must be lockable in the "off" position.
- It must be clearly labeled.
- If the inverter is located elsewhere, a second switch must also be mounted near the inverter.

The PV meter and AC disconnect switch are located outside next to the SES with clear labeling installed permanently and visible to anyone.

Courtesy of PerfectPower, Inc.

This combiner box is installed with touch-safe fuse holders and lightning arrestors for coronal discharge protection. Note the fact that this is a positively grounded systems and the system is marked to indicate that the box has been wired for positive grounding.

Courtesy of PerfectPower, Inc.

Miniature Circuit Breakers

Miniature circuit breakers provide overcurrent protection to the PV system. The PV system is automatically isolated if a short circuit or overload occurs.

Fuses and Fuse Holders

Fuses protect the wiring from overloading. Remember, string fuses must be rated for DC if used in the DC circuitry. Miniature fuses are commonly used as string fuses. You can use fuses with unearthed or ungrounded cables to prevent the module and string cables from overloading. You should locate the string fuses in the combiner/junction box. Refer to NEC 2011, article 690.16, for fuse requirements.

Ground Fault Protection

Ground fault protection monitors current flowing between the conductors and ground. When current greater than 30mA is detected, the circuit breaker will open within 0.2 seconds. Separate ground fault protection is usually not necessary because it is built into the inverter.

CHAPTER 9 SUMMARY

It is important to choose the correct type of cable for the PV application. Install cable connections correctly. Properly installed cables reduce the risk of arcing and fire. Knowing the NEC will be a great aid in choosing the right cable for your PV system design.

KEY CONCEPTS AND TERMS

Miniature circuit breakers

CHAPTER 9 ASSESSMENT

Cable Wiring and Connection Systems

1. Module and string cables differ from the AC connection cables and the DC cables.
   - ❏ **A.** True
   - ❏ **B.** False

2. Inverters are linked to the grid through the protection equipment by which kind of cable?
   - ❏ **A.** Electronic cable
   - ❏ **B.** PVC cable
   - ❏ **C.** AC connection cable
   - ❏ **D.** DC connection cable

3. Miniature circuit breakers protect the PV system from which of the following?
   - ❏ **A.** UV light
   - ❏ **B.** Array damage
   - ❏ **C.** Rodents
   - ❏ **D.** Overcurrent
   - ❏ **E.** None of the above

4. The AC disconnect switch must have which of the following attributes?
   - ❏ **A.** It must have a double pole.
   - ❏ **B.** It must be lockable in the "off" position.
   - ❏ **C.** It must be clearly labeled.
   - ❏ **D.** All of the above.

5. When maintenance needs to be performed, or if there is a short, it is important to be able to isolate the inverter from the PV system.
   - ❏ **A.** True
   - ❏ **B.** False

6. Install the DC disconnect switch before the _________________ and not in the combiner box.

❑ **A.** inverter

❑ **B.** array

❑ **C.** module

❑ **D.** grid

7. What type of device monitors the current that flows between the equipment and ground?

❑ **A.** An alternator

❑ **B.** Ground fault protection device

❑ **C.** A switcher device

❑ **D.** A generator

System Monitoring, Testing, and Troubleshooting

THIS CHAPTER EXAMINES SYSTEM monitoring, testing, and trouble-shooting, considered the last step in the design process. To finish, we also cover maintenance steps.

These factors in the PV system delivery process are critical to providing systems that work as designed and perform as predicted.

Topics & Concepts

This chapter covers the following topics and concepts:

- System monitoring
- System testing
- Troubleshooting
- Why the art and science of photovoltaics: what it all means

Goals

After completing this unit, students will be able to:

- Describe the basic principles, methodologies, and strategies for sizing photovoltaic systems to the Code and for performance.
- Evaluate system PV performance.
- List performance monitoring and testing processes.
- Create a PV system maintenance plan.

System Monitoring

Monitoring the PV system with a variety of metrics is important. Metering tells the client what the system is doing and how it is performing. Just as it is unsafe to drive a car or fly a plane without gauges, it is unwise to install a PV system where you cannot tell what the system is doing. It is important to know how well it is doing by having ready access to information. The client needs to understand what the monitoring provides and what the impact of inaction may be.

Here is the type of information that monitoring tells the client:

- Whether the system powered up in the morning and is active during the day
- What production levels are
- What faults or failures the system has experienced
- What the system's maximum and minimum power, voltage, current, temperature, and other data (depending on the inverter and monitoring system) are

With a digital graphic display, the owner can see aspects of the system's operation in a specific time period. The system may have the capability to e-mail or text notifications to the owner and integrator of any problems, irregularities, or just give a status report.

Systems are becoming more sophisticated. Some are linked to the whole house's monitoring and security systems. They can indicate when it is time for system repair and/or when maintenance needs to be performed.

Recorded metrics help avoid disputes between the client and the PV integrator while saving both of them time and money. More often than not, recorded metrics help demonstrate that what seems to be a problem really isn't one. Talking through the system over the phone with the client can reduce the need for a service visit.

When PV integrators are doing their job by monitoring the system, they can spot and repair problems before there is a loss of energy, be it small or large. Monitoring will cost less than the energy lost without it.

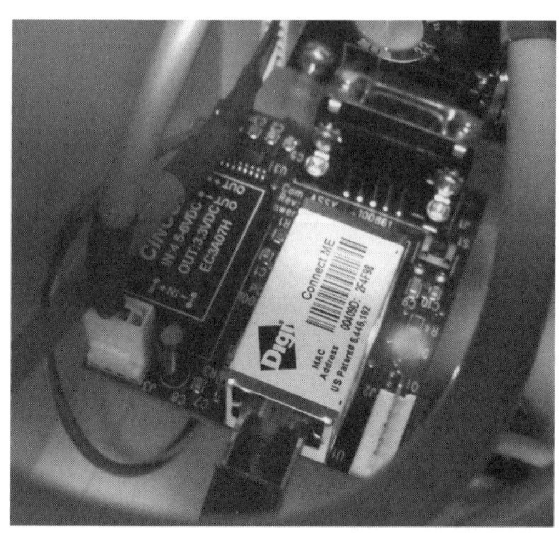

Monitoring your PV system is the same as viewing the gauges on your car and just as critical. Most inverters allow for Internet-based monitoring from any Internet-connected computer or phone anywhere around the world.

Courtesy of PerfectPower, Inc.

Guarantees

The PV designer, installer, manufacturer, or integrator can provide guarantees. Consider contractor warranties when providing guarantees. A warranty for electrical equipment is an area of concern. You need to consider roof penetrations and who will be responsible for repairs under which conditions.

Manufacturer guarantees cover work for a limited time. Guarantees are designed to protect the manufacturer, not the owner. A guarantee may cover manufacturer defects. Installer defects caused during installation will be guaranteed at the installer's expense. Installation defects include, but are not limited to, the following:

- Roof penetrations
- Electrical safety issues
- Breaches of the integrity of the structure

When your company has confidence, which it has been developing over time, it can offer a real warranty with meaning. When you design and install systems properly and check your work, whatever system problems you have tend to be minimal and to cost very little to repair. Many companies provide warranties when they have no knowledge of the technology, the quality of installation, or an understanding of which components will last far beyond the warranty period. They do not have the financial reserves to repair the system and in many instances do not have the skill to find problems and correct them.

If you protect your customer with a solid warranty, you protect your company's future. Installers need to understand their liability and the law. Many do not. It is important for the client to protect his or her investment.

Having data available at the inverter often means that it will be overlooked unless you are working on site. All PV systems need Internet-based monitoring to keep them at peak performance throughout their lives.
Courtesy of PerfectPower, Inc.

NOTE

Many jurisdictions require PV companies to provide a two-year warranty. This creates a focus on the near term, with little consideration for the long-term viability of the system.

A comprehensive PV system warranty should provide a level of performance and service that eliminates 95 to 98 percent of the problems that most clients experience. The warranty should include the following:

- A five-year, "bumper-to-bumper" warranty excluding only the monitoring system card
- Annual PV system array cleaning
- Annual detailed system-wide inspection and testing with a written report
- Weekly Internet monitoring for five years
- A maximum of 24-hour telephone response and a site visit (if needed)

You should also strive to respond to all calls for service within a maximum of 24 hours, and be prepared to make site visits if needed. Your goal is to build a sustainable business, proven successful by supporting the customers.

System Failures

Complete PV system failures are rare, but they do occur. They may be caused by the following:

- Faulty equipment
- Faulty design
- Faulty installation
- Surges from the utility side of the system
- Surges from the owner side of the system
- Lightning
- Fire
- Damage caused by animals or people
- Lack of maintenance

As the quality of design improves and standards are developed, system failures will decrease.

Typical Faults

PV systems have several common faults. They are as follows:

- Insulation faults on the DC side of the inverter can cause shocks or arcing, which can result in fires. To prevent this, do routine checks and look for damage to the insulation. Damage may be thermal or mechanical. Inverters can monitor for insulation failure as ground faults or short circuits. When sensing a failure, the inverter will isolate the PV system from the grid. Remember, the array will still be capable of sending sufficient current to cause an arc.
- Module distortion causes cells to crack or damage internal panel conductors. Time, temperature, and wind can cause cell damage. Sometimes, installation issues cause warping of the module. Be sure to select good quality modules and to properly anchor and square them during installation. Ensure anchors are made of material that will not react with the structure.
- Poor solder joints during manufacture can fail, reducing or stopping panel output.

In the past, inverters were the source of many faults in PV systems. New technology has made inverters less susceptible to damage by electrical surges and overloads. Properly testing the inverter, and recording the results, will assist in troubleshooting because you will be able to review records to see how it was functioning in the past.

Performance-Stealing Issues

Previous chapters have discussed performance issues. Good design looks for all those little incremental issues that cause a loss in energy or that block energy production.

In the PV industry, many are waiting breathlessly for the next half a percentage point of cell efficiency in the design. Many manufactures invest hundreds of millions of dollars to make ongoing minor improvements.

By addressing design and installation standards and applying the principles you have been reading about, you can make your system designs 10 percent to 50 percent more effective and higher performing than is often seen in the field. The incremental changes of a tenth or quarter percent here and a half a percent there add up quickly.

Small system losses steal system performance and output. Those losses lower system efficiency in small, often imperceptible steps. If you do not begin to look for losses during design and installation, you may never see them. Your client will be cheated right out of the gate.

Maintenance for PV Systems

Maintenance is an important part of PV systems. Checking the system on a routine basis will keep it operating at optimum levels. Monitoring and visual inspections of the system allow for timely repairs, often saving the client money.

Check the inverter or the Internet monitoring display daily when possible. Weekly monitoring will show trends, especially when you make comparisons with other systems you manage in the same area or region. Installing good quality monitoring equipment to locate faults and provide operating data will make maintenance checks easy. **TABLE 10-1** contains a chart the owner may use to inspect his or her PV system.

Array Maintenance

Array maintenance should be performed monthly or quarterly, depending upon location, condition, and performance. Check the arrays every three or four months for soil and debris. Clean arrays with a soft brush or sponge and water with mild detergent or vinegar. Never use abrasives or any chemical that will react with the aluminum frame or other metals. Take this opportunity to check the wiring connections.

Battery Maintenance

Battery maintenance varies widely depending upon the following:

- Design
- Battery type
- Application

TABLE 10-1 MAINTENANCE AND UPKEEP CHECKLIST		
Daily or weekly	Inverter	Check the display to see if the system is operating during the day. Is the system operating, showing normal performance, and operating without any faults? Note that inverters can be placed in the owner's garage for quick visual checks.
Monthly	Check kWh (instantaneous and daily yield)	Download and/or graph log readings to your computer for review. Irregularities will become apparent.
Monthly to quarterly, depending on site conditions	Module/array surface	All arrays need to be cleaned, especially of bird droppings. Most cleaning can be done with a high-pressure nozzle on a hose. For dirt, bird droppings, or oily buildup, use a mild detergent, brush, or sponge. Also inspect modules to determine if they are properly attached to the structure.
Annually	Combiner box	Check for evidence of moisture or galvanic damage. Also check for insects, debris, and animal nests.
	Surge protectors	Check the surge protector for evidence of tripping, which is a self-destructive process. Check after thunderstorms or utility outages.
	Cables, conductors, conduit, structure, and terminations	Check connections physically. Check for evidence of arcing, damage to insulation, animal ravaging, and charring. Check all mechanical and electrical connections, ensuring they are torqued to the manufacturer's specifications.
	Check for and compare against original measurements at commissioning	This is a battery of tests to be performed by a PV professional or a trained customer. Compare ratios of strings and other output data.
	Inverters	To be performed by a PV professional or trained customer.
	Modules	Test modules separately if found on a weak string. Can be performed by PV professional or a trained customer.
	Combiner box	Check the strings and fuses. Look for anything that would indicate arcing.

Cleaning the module surface is an important part of maintaining a PV system. Mild soap and water are adequate to remove dust and dirt, bird droppings, and common residues caused by air pollution. This PV system is getting its annual service cleaning with mild detergent and water. A high-pressure spray is normally used for intermediate cleanings.
Courtesy of PerfectPower, Inc.

Battery manufacturers provide recommended maintenance procedures and tasks. The warranty may be voided if these recommendations are not followed. Some batteries require more maintenance than others. All batteries require some level of monitoring and maintenance. By ignoring monitoring and maintenance, you place the batteries at risk. This can have potential safety risks.

Remember, there is no such thing as a maintenance-free battery! Maintenance tasks may include the following:

- Adding water
- Retightening terminals
- Keeping terminals clean
- Cleaning battery cases
- Keeping cables and all surfaces clean
- Checking battery performance

Check the battery containers carefully for swelling, cracks, corrosion, or electrolyte residue. To ensure the batteries are operating at full capacity, you should monitor and keep a log of the following:

- Specific gravity recordings
- Conductance readings
- Temperatures

- Cell voltage
- Results of capacity tests
- Battery voltage and current readings while batteries rest before and after charging

Many batteries emit acid and gases. A professional or a trained client should clean the batteries once or twice a year, more often if needed. Wear appropriate personal protective equipment such as gloves and goggles. Use appropriate cleaning agents for batteries. Cleaning supplies can be found at auto-supply stores. Apply battery grease to battery terminals where appropriate to protect them from acid and corrosion. Battery grease can also be found at auto-supply stores. Always follow the battery manufacturer's maintenance instructions.

Inverter and Charge Controller Maintenance

Inverters and charge controllers need to be installed with system expansion in mind. They need enough space to meet NEC requirements. They also need space so that maintenance can be performed safely and easily. Locate inverters and charge controllers in a cool, dry, and sheltered area whenever possible. Provide adequate airflow. If they are outside, keep them out of direct sunlight. This is important to help keep inverters and charge controllers working properly.

Inspect the conduit connections to the inverter box, any other switches, pass-throughs, and other conductor protection. Open the boxes carefully. Be wary of the following:

NOTE

Placing an inverter in the sun can result in 30 to 50 degree temperature increases over shaded locations.

- Insects
- Animals
- Vegetation

These can cause you or the equipment harm. If there is anything inside, carefully clean it out with the inverter isolated and off. When dealing with bees or a beehive in an inverter, have it professionally removed with the system shut down.

Inspect and torque where appropriate all terminations, again looking for any dark spots. These may be a sign of loose connections or arcing. Any discoloration should be noted and corrected.

Determine early on the percentage of variation in the inverter metering of output and the PV meter you have installed (if required). Both will have a range of accuracy. Different readings do not necessarily mean the inverter or meter is not working well.

Maintenance Tools and Equipment

Most of the tools used to install the PV system are also used to maintain the system. Some maintenance duties will require the full set of tools, and in many situations, maintenance duties will require a computer. Following is a partial list of tools that maintenance personnel will need to perform maintenance duties on PV systems:

- Multimeter electrical testing instruments (these need to be able to test up to 600V AC/DC resistance and continuity, unless you are working on higher-voltage equipment)
- Gloves
- Torque wrench and torque screwdrivers
- Basic hand and power tools
- Ladders
- Personal protective equipment (PPE)
- Solar shading devices
- A **pyranometer** or preferably a **pyroheliometer**. These are two types of devices that measure the intensity of sunlight
- Hydrometers, which check the specific gravity of the electrolyte in a flooded lead-acid battery
- Standard digital camera capable of 300 dpi and an infrared camera or thermometer (the camera is your best source for seeing and recording the range of temperatures, and the thermometer enables you to measure temperatures of components without having to touch the equipment)
- Insulation and ground resistance testers

Handheld devices or personal computers enable maintenance personnel to monitor system performance in real time, download historical data, process string voltages, and program and control system functions.

System Testing

System testing is important to the life and output of the PV system. Many inverters make comprehensive monitoring easy for the client to perform. Most modern inverters feature monitoring in one of the following forms:

- Integrated into the inverter
- As an add-on remote display
- As an Internet card that allows access to the monitoring data via the Internet

The PV integrators should include in the contract the right to access the monitoring data to stay on top of any issues or challenges. In some monitoring systems, alarms and e-mail notification are included or can be added to the system to inform the client and integrator of any problems.

Remote monitoring software simplifies troubleshooting, provides additional information, and provides frequent readings of the system. It refreshes at various rates, usually in one- to 15-minute intervals.

Some systems include climatological kits, which include the following:

- A pyroheliometer or pyranometer
- An anemometer for wind speed
- Temperature tools
- Wind-direction tools

Irradiance sensors measure actual conditions at the site. These tools can be quite helpful for monitoring, testing, and troubleshooting. They are more common on larger systems. This data compares the system's anticipated output to the actual output. Calibration can be automatic, and helps in troubleshooting problems. Some monitoring systems use the Internet to e-mail the PV professional when a number of faults or voltage issues occur. This allows for a quick diagnosis and repair if needed.

Post-Installation Testing

For the highest level of reliability and performance, take into consideration the 90–100 day post-installation burn-in phase, during which equipment is in standard operation, electrical and mechanical equipment and connections have an opportunity to expand and contract, and minor shifts in equipment take place as they properly seat.

Burn-in testing is the same level of testing that occurs in the post-install, pre-commissioning phase, where the technician goes through the whole system with a checklist and reviews everything.

Why perform a post-install burn-in test when it takes time and money? Once again, it is the little things that eat away at performance. Finding the loose conductor that bleeds off a little energy, the terminal or mechanical fastener that has loosened through expansion and contraction, or the bad solder in a panel that now affects the string may eliminate one or more service calls in the future.

There is more to it than that, however. This phase represents an ongoing training opportunity and a chance to improve quality control. And of course, your customer has probably never had anyone come back without an uncomfortable phone call. If you are going to be a quality PV organization, you have to provide the highest quality service.

A well-trained and well-organized service tech can do an exceptional system test in a relatively short time.

NOTE

There are two times a year when the PV system's output will peak. They are the spring and fall equinoxes.

Troubleshooting

When properly designed and installed, PV arrays can have a lifespan of 40 years or more. As with any product, problems, faults, and failures can develop during this time. If the integrators and the customer are monitoring the system, an irregularity will usually be discovered soon after the issue arises. Unfortunately, however, that is not how most customers and integrators deal with monitoring. Many see it as an added expense, not as a valuable tool.

As someone who wishes to pursue a career in the PV industry, you must understand that a good portion of your success will be determined by how well you can find and resolve problems in systems.

Following are several troubleshooting processes. Take system measurements and verify they are the similar to your commissioning measurements. Visually check all wiring, conductors, conduit, and terminations. Be sure to follow all appropriate safety protocols while troubleshooting. After gathering information from the client, perform the following:

- Confirm as-builts.
- Check for shading.
- Check for loose wiring connections.
- Check for grounding issues and insulation-related faults. Test strings individually and compare output for mismatches.
- Check for short-circuiting caused by faulty insulation.
- Check for faults at the inverter. Data from the inverter will help locate any problems.
- Look for defective string fuses. Use a voltage meter to measure differences in strings.
- Check for module defects. Use an infrared camera to view panels and strings.
- Check for surge-protector defects or failure.
- Check for any and all visual irregularities.

Monitoring Operating Data and Presentation

Monitoring the PV system is an essential step to keeping it operating at commissioning levels. Inverters play a large role in a comprehensive monitoring system. Inverters collect system and operating data for the client. The data can be read from the inverter's screen/display. The inverter can also send the monitoring data to a computer anywhere in the world that has Internet service. There, it can be reviewed. The monitoring data can also be sent via computer to a service company that will interpret the data as a separate contract item.

Comparing and analyzing data makes it possible for the client to know that the system is working properly. Constant monitoring tells the client when the system has begun to develop a problem or when a fault or failure has taken place. Weather, insolation, voltage, electrical currents, and all other system variables change all the time. Being able to see output data visually is incredibly valuable. Monitoring the PV system enables the client to determine whether the system is operating at optimum levels, which justifies their investment.

Internet-Based System Evaluation

An Internet-based system sends data to a remotely located system operator, either with the inverter manufacturer or contracted by that organization. Data is collected and then stored on a server. The server processes and analyzes the data. Graphs and charts can be made available for the client to view at his or her discretion. Clients usually access this data through a Web site using a confidential password.

Manufacturers hold the data for different periods of time. It may be important to determine what their retention period is to pass on to the customer.

Most PV organizations do not realize they can use this data to present a performance picture to new customers. This will help to explain how well the systems they install and maintain will operate. Showing actual operating data to new clients can be invaluable.

Web-Based Data Transmission and Evaluation

The Internet is evolving into every aspect of our lives. PV systems are no exception. The Internet allows PV system data to be transmitted and in some instances translated into a wide range of computer languages. This makes the data readable by different programs, which are entering the market at an incredible pace.

The International Electrical Code 62350 Section 5.4 proposes that there be a standardization of terms for PV systems. Standardization of terms will make it easier for companies to collect and compare data. This will allow the PV industry to evolve and create systems that are more efficient.

Presentation and Visualization

Being able to visualize the operating data is becoming important to clients. Sometimes, it is only about bragging rights. However, many customers want to know they got their money's worth. Being able to see the monitoring data graphed on a computer screen helps the client understand how the system is operating. This also enables the client to show the benefits of the PV system to friends and family without having to climb up onto the roof!

Long-Term Behavior of PV Modules

Good-quality PV modules can last and perform for a long time. Some crystalline modules have been operating for 50 years. They have experienced reduced efficiency over time, but not so much as to require their replacement. Many PV panels are built to last and not degrade, and to produce more energy per DC watt. Carefully installed modules made of high-quality cells with exceptional quality assurance in components and workmanship will provide high output for decades.

Weather, heat, and ultraviolet light age modules. Signs of aging are bleaching, brown spots, and corrosion. Different products and technologies have different

PV modules are the backbone of any system. It is important to select high-quality modules that can endure decades of use and wear.
Courtesy of PerfectPower, Inc.

overall derating schedules and challenges. Some products will last less than a decade, and some far longer as mentioned. It is your responsibility to determine which products last and which ones do not. This creates a real dilemma. You have to ask yourself, "Do we go with the cheap products and take a risk for our company and customers, or do we rely on products with a solid track record?" Some solar energy firms have made great choices in that vein, while others have not. Many are no longer viable companies because of bad technology choices.

Plastic-coated cells can delaminate, get brittle, crack, and short out. This tends to occur in areas with high humidity, high altitudes, high UV content, rapid temperature changes, and high temperatures. This creates cell corrosion, discoloration, and bleaching. Output decreases far more rapidly than the approximately 0.20 percent per year claimed by some manufacturers for their products.

Modules with thin-film technologies may degrade more quickly than many other types of modules. During the first 1,000 hours of operation, they can lose 10 to 15 percent of their output. Thin film reliability is still being proven in the industry and will in time do much better.

Quality and Reliability of Inverters

Inverters are the most sensitive part of the PV system. In most systems designed only to code, they tend to have an average life of approximately 10 years or less. Inverters must be treated delicately to live a long life. Repairs or complete replacement will need to be performed based on how they are designed into the system, the environmental conditions in which they are placed, and a number of other factors we have discussed. Service agreements from manufacturers are available, as are 20-year extended guarantees for an additional price.

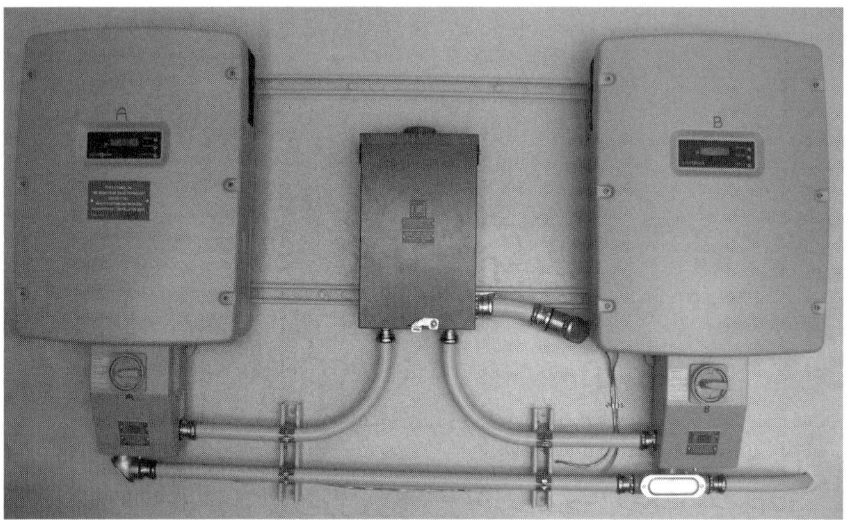

This set of inverters has been installed in a location with some temperature control to protect it for a life of up to 20 years.
Courtesy of PerfectPower, Inc.

Large output losses are usually caused when the client does not discover a faulty inverter. If the inverter is not performing properly, potential energy collection is wasted. Kilowatt hours (kWhs) are money in your pocket, or not. These are often avoidable with proper design and installation, and with the use of a monitoring program. Using surge protectors increases the reliability of the inverter.

Why the Art and Science of Photovoltaics: What Does It All Mean?

Photovoltaic (PV) technology is about 175 years old!

What are we attempting to do with this series of texts and what will it take to propel the PV industry into the next century? The technologies that are gathered under the PV umbrella of technologies are diverse. Yet there are many myths about the technologies and their applications.

This educational program is designed to separate myth from reality while providing you the most advanced design and installation knowledge techniques and the underpinning standards necessary to support them.

This text is designed to give you a practical and structured understanding of the PV technologies, their strengths, and their weaknesses, as well as how to apply the technologies in a manner that will harvest the greatest energy yield while achieving the greatest cost effectiveness.

Good PV design and installation is both an art and a science. It's an art because there are numerous ways to apply the technologies. This requires critical thinking, planning skills, and creativity to achieve the safest, most aesthetically pleasing and customer-friendly applications. A PV system can be visually well designed and installed, or it can look like someone's grade-school science project. Customers care about how their PV system looks almost as much as they care about how well it operates.

PV is also a science. Your goal is to properly apply the technology. Yet most people do not take the time to learn or understand its application. As a result, most have not developed the knowledge and skills to maximize PV technology capabilities. Do you believe that knowledge is power? In the PV world, power is nothing until it's properly applied. The same is true of knowledge and getting the most out of the technology!

Understanding all of the components and how they work together allows you to produce 10 to 30 percent more energy as you properly integrate these components. You can be a C- or D-class solar professional. Alternatively, by applying the techniques and understanding these technologies, you can become an A-class professional integrator if you choose to do so. It's your choice!

Photovoltaic integration is about the energy business. It is always about the money. To be a sustainable industry, PV systems need to produce energy reliably and productively, and do so for 20 to 40 years with minimal operations and maintenance costs. This requires that students develop the ability to see PV as the design and installation of a complete system, not the simple commingling of a bunch of low-cost parts.

Students will learn PV systems design theory, which gives them an advantage in the way they put systems together and operate. It is also a great opportunity for you to gain a serious professional advantage in the work and marketplace.

You should strive to break the pattern of using the National Electrical Code (NEC) as the defining design and installation tool and move on to the 36 PV system evaluation, design, and installation criteria groups. If you can make this transition, it will help you to raise the standards bar. NEC is the safety foundation, not the ultimate product or goal. It is the minimum legal standard for safety. It provides an incredibly important set of tools. Yet, when designing and installing, everyone in the industry should learn to do better.

In this process, we will define three levels of PV system delivery. These include subprime, moderate, and high-performance. System buyers have a right to know what they are buying. They need a simple and understandable way to determine if they are receiving what they think they purchased. To be happy, they must receive value. As a result, we are providing the language and metrics, the yardstick, to make good financial and technology decisions. In the process, we give you the tools to deliver moderate and high-performance systems.

The process we have embarked upon is to link the evaluation, design, and installation chain in a coherent and productive manner so that:

- Systems are designed to meet customer needs.
- Technologies and components are selected to meet design and energy-yield goals and objectives.
- Systems are designed to be reliable, robust, and productive.
- Systems are designed to be installed in a more effective manner.
- Installation becomes the physical manifestation of the PV system design, production, operation, and customer goals.
- Installation takes the materials and methods through a logical quality-assurance process to reduce performance challenges, service calls, and O&M costs.
- Systems are designed and installed to be tested and monitored to ensure better operations, maintenance, service, and usability life.

For the PV industry to be sustainable and vital, continuous improvement must be the standard.

If you are seeking a career in the PV industry, this series is designed to give you the basic and initial advanced skills and knowledge to make you a higher-performing solar employee or employer. If you are educating yourself to make corporate decisions to purchase PV technologies for your own premises, this series will help you develop the tools to enable your organization make better, more productive decisions.

The information in these books took over three decades to develop. The goal of these books is to reduce the basic learning curve for the student who wants to excel in the PV industry. Solar energy technology is dynamic, exciting, and almost magical in the way it converts abundant and free sunlight into usable energy. My hope is that you will take the information we are presenting and use it to lead in the PV industry.

You are embarking on a technology path, and your learning should end only when you are no longer working in the industry. Unfortunately, many people quit learning after a few years in the industry. Please keep in mind the need to always be on the PV learning path.

CHAPTER 10 SUMMARY

PV system monitoring is important to the life of the system and the customer's peace of mind. The inverter is an important part of collecting system data. Data can be sent by the Internet to a company that will analyze the information if the owner wants professional monitoring. The company will inform the client of how the system is performing.

Problems will be identified and timely repairs can be made. Keeping the system free from faults will keep it operating at the optimum level. This will keep the client happy.

KEY CONCEPTS AND TERMS

Internet-based system
Pyranometer
Pyroheliometer

CHAPTER 10 ASSESSMENT

System Monitoring, Testing, and Troubleshooting

1. Heat lowers PV module output.
   - ❏ **A.** True
   - ❏ **B.** False

2. When is system output at its highest?
   - ❏ **A.** Sunny days in spring around the equinox
   - ❏ **B.** Sunny days in summer around the solstice
   - ❏ **C.** Sunny days in fall around the equinox
   - ❏ **D.** Sunny days in winter around the solstice
   - ❏ **E.** A and C
   - ❏ **F.** B and D

3. PV installers often guarantee what aspects of the PV system?
   - ❏ **A.** Roof penetrations
   - ❏ **B.** Electrical safety issues
   - ❏ **C.** Structure integrity
   - ❏ **D.** All of the above
   - ❏ **E.** None of the above; they are covered by manufacturers

4. What do meters and gauges do?
   - ❏ **A.** They provide information to help avoid disputes between the client and the PV installer.
   - ❏ **B.** They control voltage.
   - ❏ **C.** They control current.
   - ❏ **D.** They convert DC to AC.

5. What does post-commissioning burn-in testing provide?
   - ❑ **A.** A longer life for the PV modules
   - ❑ **B.** An opportunity to find system faults that occur normally
   - ❑ **C.** An opportunity to fine-tune the system
   - ❑ **D.** An opportunity to avoid potential service calls and energy loss
   - ❑ **E.** B, C, and D

6. What should the PV system owner do daily or weekly?
   - ❑ **A.** Check the inverter display.
   - ❑ **B.** Clean the array.
   - ❑ **C.** Look for corrosion.
   - ❑ **D.** Compare daily output to the day's irradiance level.

7. You should perform full system maintenance about once a year.
   - ❑ **A.** True
   - ❑ **B.** False

8. Use a _______________ to remove debris that is caked on the array.
   - ❑ **A.** bristle brush
   - ❑ **B.** power washer
   - ❑ **C.** sponge or brush
   - ❑ **D.** abrasive cleaner
   - ❑ **E.** A, B, and C

9. Many inverters feature _______________ to alert owners to problems.
   - ❑ **A.** e-mail service
   - ❑ **B.** automated phone calls
   - ❑ **C.** e-mail fault and failure alarms
   - ❑ **D.** None of the above

10. Inverters in systems based only on Code have an average life of about _______________ years.
    - ❑ **A.** 5
    - ❑ **B.** 10
    - ❑ **C.** 15
    - ❑ **D.** 20

Answer Key

Chapter 1   Overview of Advanced Photovoltaic (PV) System Design and Design Criteria

1. B, 2. A and D, 3. B and C, 4. B and D, 5. A, 6. B, 7. A, 8. C, 9. B and C, 10. E.

Chapter 2   Evaluation and Design Criteria Part II

1. A, 2. A and C, 3. B and C, 4. D, 5. D, 6. B, 7. A, 8. D, 9. B and C, 10. A.

Chapter 3   Determining the Size of a Photovoltaic System

1. B, 2. D, 3. B and C, 4. B, C, and E; 5. D, 6. A, 7. B, 8. A, B, and C, 9. B, 10. A.

Chapter 4   Photovoltaic Array Configuration and Sizing

1. A, 2. D, 3. A and D, 4. B, C, and E; 5. D, 6. A, 7. A, 8. D, 9. A, B, and C; 10. A

Chapter 5   Mounting Systems

1. A, 2. B and D, 3. A, 4. A, 5. A, 6. A, 7. A, 8. A, 9. B and C, 10. B, C, and E

Chapter 6   Energy-Storage Device (ESD) System Design

1. B, 2. A, B, and C; 3. B and C, 4. B, 5. D, 6. A, 7. A, 8. D, 9. A, 10. A, B, and C

Chapter 7   Charge Control and Maximum Power Point Tracking (MPPT)

1. A, **2.** F, **3.** B, **4.** B, **5.** B, **6.** B, **7.** D, **8.** D, **9.** B, **10.** C, **11.** B and C, **12.** B, D, and E

Chapter 8   Inverters in System Design

1. B, **2.** C, **3.** B and C, **4.** B, C, and E; **5.** E, **6.** A, **7.** A, **8.** A, D, and E; **9.** B and C, **10.** B, C, and E

Chapter 9   Cable Wiring and Connection Systems

1. A, **2.** C, **3.** D, **4.** D, **5.** A, **6.** A, **7.** B

Chapter 10 System Monitoring, Testing, and Troubleshooting

1. A, **2.** E, **3.** D, **4.** A, **5.** E, **6.** A, **7.** A, **8.** E, **9.** C, **10.** B

Glossary

Alternating current (AC)   An electrical current that reverses direction.

Ampere (A or amp)   A unit of electric current that is the rate of the flow of electrons in a conductor, equal to 1 coulomb per second.

Ampere-hour (Ah)   Used to identify energy stored in a battery. The amount of electrical energy that equals the flow of current of 1 ampere for 1 hour.

Array   A set of electrically connected photovoltaic (PV) modules.

Array-reconnect voltage (ARV) set point   The ARV set point is where the array stops charging the battery and battery voltage drops.

Array-to-load energy ratio (A:L)   The ratio that represents the average daily usable amount of energy output by the array.

Battery   Converts chemical energy into electrical energy by means of an electrochemical oxidation-reduction (redox) reaction.

Battery capacity   The total number of ampere-hours that are available from a fully charged battery at a constant current flow.

Battery cell   The part of a battery that stores electrical energy for use by an external load.

Battery cycle life   The number of times a battery can be fully discharged and recharged before falling below the 80-percent capacity charge maximum at the end of its life.

Battery self-discharge   Energy loss by a battery when not under load.

Charge controller   A device that controls the rate and/or state at which batteries receive charge from the modules.

Clipping   The DC energy loss that occurs because oversized and undersized arrays are being matched with improperly sized inverters and or they are sized outside the MPPT performance window.

Crystalline silicon   A single crystal or polycrystalline slice of silicon used to make a PV cell.

Cycle   The discharging and subsequent charging of a battery.

Deep-cycle battery   A heavy-duty battery that can withstand many deep discharges.

Diode   An electronic component that allows current flow in one direction only.

Direct current (DC)   An electrical current that moves in one direction.

Discharge   The withdrawal of electrical energy from a battery.

Duty cycle   The period when a device is operating, using energy to function. For example, when a refrigerator motor is running to cool the inside, the refrigerator is in the duty cycle.

Duty rating   The amount of time the PCU can operate at the maximum load.

Frequency   The load frequency in cycles per second is required in selecting a PCU. In the United States, most loads require 60Hz, whereas in other parts of the world, 50Hz is commonly used. Variations in frequency can cause poor performance of clocks and electronic timers within the system.

Input voltage   Determined by the PCU/inverter design characteristics to power the required AC loads. The input voltage defines the array voltages required for effective operation of the inverter.

Inverter   Also known as a power conditioning unit (PCU) or a power conditioning system (PCS), an inverter converts DC power from the PV array or battery bank to AC power that is compatible with the utility and AC loads.

Internet-based system   This type of system sends data via the Internet by a modem to a remotely located systems operator, either with the inverter manufacturer or controlled by them under contract.

I-V curve   The curve of the electrical current versus voltage characteristics of a photovoltaic cell, module, or array. The three points on the I-V curve are the open-circuit voltage, the short-circuit current, and the peak power operating point.

Lead–acid battery   A category of batteries with plates made of lead that are immersed in an acid electrolyte.

Load profile   The variation in the electrical load in watts versus time (kilowatt-hours) based on a comprehensive list of devices that require electricity to operate, provided by the client.

Load-reconnect voltage (LRV) set point   The LRV set point is the voltage value where the load is reconnected to the battery.

Load shifting   An energy-management tool and technique used to move demand from the peak hours to off-peak hours of the day. For example, using a PV system, wind turbine, hydro turbine, or generator to power a portion of a load reduces the peak hour cost of electricity. This is most often accomplished when the load is shifted to an energy generator that costs less to produce a kilowatt, thereby creating savings.

Load-transfer switch   A device used to load a functioning inverter in the event an inverter fails in a PV system with multiple inverters.

Low-voltage disconnect (LVD) set point   The voltage at which the charge controller will disconnect the load from the batteries to prevent overdischarging.

Low-voltage disconnect hysteresis (LVDH)   The voltage difference between the low-voltage disconnect set point and the voltage at which the load will be reconnected.

Maximum power point tracking (MPPT)   When the array operates at the peak power point of the array's IV curve, which is where maximum power is obtained.

Maximum system voltage   A politically determined maximum voltage allowable by code.

Miniature circuit breakers   These provide overcurrent protection to the PV system.

Module   A unit made of PV cells.

Modularity   Having multiple PCUs in a PV system. These can be connected in parallel or used to service different loads. Modularity is also becoming a greater part of inverter design, where inverter components are modular and can be replaced or rebuilt to extend the life of the inverter. This in turn reduces its usable life cost. Modularity is an important design decision when selecting the right inverter.

Nickel-cadmium (NiCd) battery   A battery that contains nickel and cadmium plates in an alkaline electrolyte.

Panel   PV modules assembled in a frame.

Parallel-operation capability   In PV systems with multiple inverters, you can wire the inverters in parallel to enable them to service more load simultaneously. This is referred to as parallel-operation capability.

Peak power   The tested power output at STC. This is used as the panel or nameplate power rating.

Performance test conditions (PTC)   PVUSA or performance test conditions (PTC) are defined as 1000 W/m^2 plane-of-array irradiance, 20 degrees Celsius (68 degrees Fahrenheit), ambient temperature, and 1 m/s wind speed. PTC differs from standard test conditions (STC) in that PTC of ambient temperature and wind speed result in a cell temperature of about 50 degrees Celsius (122 degrees Fahrenheit) instead of the 25 degrees Celsius (77 degrees Fahrenheit) for STC. PTC was developed to simulate real-world conditions.

Phantom load   The energy an electrical appliance consumes while turned off or in standby mode.

Photovoltaic cell   A semiconductor material used to convert solar irradiance into electricity.

Power conditioning unit (PCU)   Another term for inverter.

Power conversion efficiency   The ratio of output power to input power of the PCU. The efficiency of a standalone PCU varies significantly with the type and amount of load.

Power factor   The cosine of the angle between the current and voltage waveforms produced by the PCU. This value varies with the load. The power factor should be very close to 1.

Pyranometer   A device that measures the intensity of sunlight.

Pyroheliometer   A device that measures the intensity of sunlight.

Rated current   The current output of a PV module measured at standard test conditions (STC) of 1,000w/m^2, 25 degrees Celsius cell temperature, and at 1.5 atmospheres.

Rated power   How many watts of power the PCU can supply during standard/nominal operation.

Remote control/data monitoring   Allows the inverter to be controlled or monitored remotely.

Sealed battery   A sealed battery has a resealing vent cap that releases gas and moisture under pressure. The electrolyte is captive and cannot be replenished. Also known as a valve-regulated lead-acid (VRLA) battery.

Series controller   A controller that interrupts the charging current by open-circuiting the PV array. The control element is in series with the PV array and battery.

Short-circuit current   The current is at the maximum voltage where the panel is shorted out at STC.

Shunt controller   A controller that redirects or shunts the charging current away from the battery. The controller requires a large heat sink to dissipate the current from the short-circuited PV array. Most shunt controllers are for smaller systems producing 30 amperes or less.

Single-stage controller   A unit that redirects all charging current as the battery nears full SOC.

State of charge (SOC)   A measure of how full of energy a battery is, stated as a percentage of 100 percent charge minus the depth of discharge.

Surge capacity   The capability of a PCU to exceed its rated power for a limited period, usually in seconds.

Surge load   The electrical demand when a device is started that exceeds the device's nominal operating electrical demand.

Voltage protection   Sensing circuits that disconnect the unit from the battery, array, or inverter if either the voltage or current is too high.

Voltage regulation   Indicates the variability in the input voltage for batteries or the output voltage for inverters. Better PCUs produce a nearly constant root-mean-square (RMS) output voltage over a range of loads.

Voltage-regulation hysteresis (VRH)   The voltage difference in the form of a lag between the VR and the AVR.

Voltage-regulation (VR) set point   The maximum voltage that the battery is permitted to reach at a specific temperature.

Watt (W)   A unit of electrical power that occurs when a current of 1 ampere flows through a conductor with a potential 1 volt. A single watt equals 1/746 of a horsepower.

Watt-hour (Wh)   An energy unit of 1 watt of power operating for 1 hour, usually expressed in units of 1,000 watt-hours, called kilowatt-hours (kW h).

References

Australian Business Council for Sustainable Energy. (n.d.) "Electricity from the sun: Solar PV systems explained." Carlton, Victoria, Australia. Retrieved June 7, 2001, from http://www.ntca.org.au/_assets/Solar_Electricity.pdf.

Burger, B. & D. Kranzer. (n.d.) "Transformerless PV topologies." Freiburg, Germany: Fraunhofer Institute for Solar Energy Systems. Retrieved May 5, 2011, from http://powersystemsdesign.com/index.php?option=com_content&view=article&id=453&Itemid=83http://powersystemsdesign.com/index.php?option=com_content&view=article&id=453&Itemid=83.

Chiras, D. (2009). *Power from the sun.* Gabriola Island, British Columbia: New Society Publishers.

Dunlop, J.P. (1997)."Batteries and charge control in stand-alone photovoltaic systems fundamentals and application." Albuquerque, NM: Sandia National Laboratories.

Dunlop, J., & B. Farhi. (2001). "Recommendations for maximizing battery life in photovoltaic systems, a review of lessons learned." Orlando, FL: Florida Solar Energy Commission (FSEC), University of Central Florida.

Durability & Design: The Journal of Architectural Coatings. (2010, Dec. 13). "Benjamin Moore flips switch to solar installation at R&D site." Retrieved May 5, 2011, from http://www.durabilityanddesign.com/news/?fuseaction=view&id=4772&nl_versionid=707&trackid=21478787.

Endecon Engineering with Regional Economic Research, Inc. (2001). *A guide to photovoltaic (PV) system design and installation.* Sacramento, CA: California Energy Commission.

The German Solar Energy Society. (2008). *Planning and installing photovoltaic systems: A guide for installers, architects and engineers.* London: Earthscan Publishers.

Gevorkian, P. (2007). *Solar power in building design: The engineer's complete project resource.* New York, NY: McGraw-Hill Professional.

King, D.L., W.E. Boyson, & J.A. Kratochvil. (2004). "Photovoltaic array performance model." Albuquerque, NM: Sandia National Laboratories.

National Fire Protection Association. (2010). *NFPA 70® National Electrical Code® 2011 Edition.* NFPA: Quincy, MA.

National Renewable Energy Laboratory. (2006). "A review of PV inverter technology cost and performance projections." Golden, CO. Retrieved April 24, 2011, from http://www.nrel.gov/pv/pdfs/38771.pdf.

North American Board of Certified Energy Practitioners. (April 2009).

Study guide for photovoltaic system installers, version 4.2. Retrieved May 2, 2011, from http://www.nabcep.org/wp-content/uploads/2008/11/nabcepstudyguidev4-2april2009.pdf.

Solar Energy International. (2004). *Photovoltaics design and installation manual: Renewable energy education for a sustainable future.* Gabriola Island, British Columbia: New Society Publishers, 2004.

Spries, D., & R. Royer. (1998). "Guidelines for use of batteries in photovoltaic systems." Natural Resources Canada, Canadian Energy Diversification Laboratory of CANMET (CEDRL), NETE Advanced Power Systems (NAPS), Finland.

Wenham, S. R., M. A. Green, M. E. Watt, & R. Corkish. (2007). *Applied photovoltaics, 2nd Edition.* London: Earthscan Publications Limited.

Wiles, J. (2000, June/July). "Where to use which conductor." *Home Power* No. 77.

Wiles, J. (2007, July/August). "PV wiring: Continuous currents through curious cables." *IAEI News.*

Wiles, J. (2005, February). "Photovoltaic power systems and the 2005 National Electrical Code: Suggested practices." Retrieved June 7, 2011, from http://www.altestore.com/store/media/pdfs/photovoltaic_NEC_code_practices2005.pdf.

Index

Figures and tables are indicated by *f* and *t* following the page number.